Space

In the 1960s the rivalry between two great powers took humans into space, adding a new dimension to life on Earth. The final frontier was unlocked, the solar system appeared to be within our grasp. Next stop: Mars. But then nothing. The space race had declared its winner and revealed its true nature: a competition driven by patriotic and ideological pride. Between 1969 and 1972 twelve men – no women – walked on the moon, and then no more. The space agencies have not been sitting idle all this time, however. In spite of lower budgets and different objectives, we have continued to explore the solar system with probes and rockets, which is cheaper than sending humans and allows us to observe the celestial bodies from closer quarters. With politics removed from the equation, science has been the winner. But the squeezing of government funds has brought about another transformation, a textbook case of opening up a market to the forces of capitalism: NASA and other space agencies have been forced to become ever more reliant on private companies to build modules and rockets, and a generation of visionary, megalomaniacal entrepreneurs is determined to take us back into space, this time to stay. For the project to succeed, they believe it is essential to reduce costs and to exploit the resources we find up there. The race is back on, but the rules have changed, and with the rise of China and India and the emergence of middle-ranking powers in an increasingly multipolar world, the players have changed as well. But for those earthlings like us who are not involved in the race, space offers something else: a spiritual dimension that is only reinforced by science in its quest for answers to ancient questions – what is the universe made of, how was the solar system formed and how did life come about? – questions that we are no longer so accustomed to asking ourselves, guided as we are in our terrestrial wanderings not by the starry sky but by GPS. And while we know that colonising Mars will not solve our problems, the promise of space – whether conveyed by one of Elon Musk's tweets or a photograph taken by a NASA rover on the red planet – is that we will finally prove ourselves to be the intelligent life forms that can, if required, work together to build a shared future for the whole of humanity.

Contents

Page 2: For the first time in eight months, researcher
Jocelyn Dunn takes a walk outside without a helmet
(although still wearing a space suit). She had been
locked up with five colleagues in a dome in the crater
of Mauna Loa, the volcano on Big Island, Hawaii,
to simulate the isolation and living conditions that
might be experienced by astronauts on missions to
Mars. Credit: Gaia Squarci
Right: Rocket debris, labelled 'Progress', in the
garden of a house in the village of Dolgoshechlye in
the border security zone within the Mezensky District
of northern Russia. Credit: Raffaele Petralla

Some Numbers

Active satellites in orbit 1957–2021

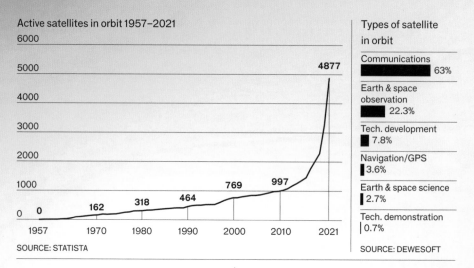

SOURCE: STATISTA

Types of satellite in orbit

Type	
Communications	63%
Earth & space observation	22.3%
Tech. development	7.8%
Navigation/GPS	3.6%
Earth & space science	2.7%
Tech. demonstration	0.7%

SOURCE: DEWESOFT

Proportion of military and civil expenditures of the 10 largest space programmes (2018)

■ Military ☐ Civil

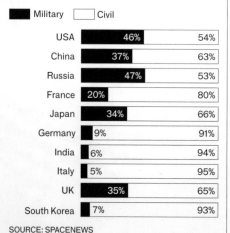

	Military	Civil
USA	46%	54%
China	37%	63%
Russia	47%	53%
France	20%	80%
Japan	34%	66%
Germany	9%	91%
India	6%	94%
Italy	5%	95%
UK	35%	65%
South Korea	7%	93%

SOURCE: SPACENEWS

Estimated cost per astronaut, adjusted for inflation (selected craft, 2019)

millions of $

/// Reusable   ■ Single use   *Experimental

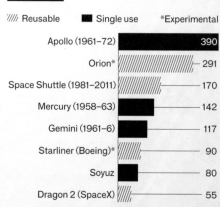

Apollo (1961–72)	390
Orion*	291
Space Shuttle (1981–2011)	170
Mercury (1958–63)	142
Gemini (1961–6)	117
Starliner (Boeing)*	90
Soyuz	80
Dragon 2 (SpaceX)	55

SOURCE: STATISTA & THE ECONOMIST

A SPACE OF ONE'S OWN

As at the end of 2021 only 596 people had travelled into space. Of those, 72 (*c.* 12%) have been women.

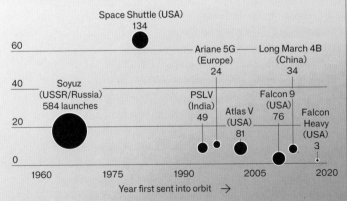

SPACE SWEET SPACE

Longest space missions, in days

1994–5
V. Polyakov (Russia)
437.75

1998–9
S. Avdeev (Russia)
379.62

1987–8
M. Manarov
and V. Titov (USSR)
365.94

2015–16
M. Kornienko (Russia)
and S. Kelly (USA)
340.36

THE LONELINESS OF THE LONG-DISTANCE ROVER

45.16

km travelled by the Mars rover *Opportunity* in 2004–18, the furthest by any rover

VERTIGO

The highest mountains in the solar system from base to summit

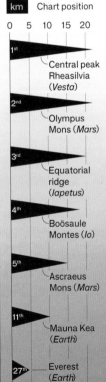

km Chart position

0 5 10 15 20

1st — Central peak Rheasilvia (*Vesta*)

2nd — Olympus Mons (*Mars*)

3rd — Equatorial ridge (*Iapetus*)

4th — Boösaule Montes (*Io*)

5th — Ascraeus Mons (*Mars*)

11th — Mauna Kea (*Earth*)

27th — Everest (*Earth*)

SPACE CARGO

Estimated cost in $ (adjusted for inflation) per kg of cargo to achieve low Earth orbit (selected craft, 2021)

↑ Thousands of $/kg

Space Shuttle (USA)
134

Ariane 5G (Europe)
24

Long March 4B (China)
34

Soyuz (USSR/Russia)
584 launches

PSLV (India)
49

Atlas V (USA)
81

Falcon 9 (USA)
76

Falcon Heavy (USA)
3

60

40

20

0

1960   1975   1990   2005   2020

Year first sent into orbit →

A SCIENCE
IS BORN

A meteorite from Mars, a planet that shouldn't exist, life forms that thrive in extreme conditions ... In the space of a few decades a series of improbable, thrilling discoveries has transformed research into one of the greatest mysteries to puzzle humankind into a genuine branch of science. Astrobiology brings together different disciplines in an attempt to answer the big question: is there anybody out there?

JO MARCHANT

9

JO MARCHANT is a science writer and journalist with a PhD in microbiology. She has worked as an editor for *Nature* and *New Scientist,* and her articles have appeared in *The New York Times*, the *Guardian* and *Smithsonian* magazine. Her works, including the bestselling *Cure: A Journey into the Science of Mind Over Body* (Crown, 2017, USA / Canongate, 2016, UK), explore the nature of humanity and how we have attempted to make sense of our place in the universe. Her latest book, *The Human Cosmos*, from which this article is taken, was published in 2020 by Dutton in the USA and Canongate in the UK.

FRANCESCO MERLINI is an Italian photographer who undertakes long-term personal projects alongside his work as a press and editorial photographer. His photos have been published in magazines and newspapers internationally and exhibited all over the world in both group and solo exhibitions. In 2020 he was nominated for the Prix HSBC pour la Photographie. In 2021 his first book, *The Flood*, was published by Void.

'The Last Shuttle: The End of an Era' is a series Merlini shot in a valley in the Italian Alps. Drawing on the magical-realist tradition, he reflects on the relationship between place, memory and mourning, creating a visual path in which typical elements of the mountain panorama acquire new meanings and become archetypes of memory.

December 27 1984 was a warm summer's day in Victoria Land, Eastern Antarctica, with a lull in the normally biting winds and the temperature hovering as high as –20 degrees Celsius. At the foot of the Transantarctic Mountains the ice forms a vast flat plain that dazzles in the sunlight and is tinted slightly blue. Geologist Roberta Score and her colleagues spent the morning patrolling the area on snowmobiles, sweeping back and forth for hours in a search formation around thirty metres apart. The monotonous routine could numb the eyes and brain, so just before noon the team leader, John Schutt, called a break, and the group diverted to a nearby escarpment surrounded by ice pinnacles, carved by the wind, that look like giant, frozen waves.

After enjoying the view from the top of the ridge, the geologists rode back towards their search area, taking care to avoid chasms in the ice and windswept piles of snow. And that's when Score saw it: a dark-green spot against the glaring blue. She stopped her snowmobile and waved her arms to alert the others. Here, on this ancient untouched stretch of ice, was what they had been looking for: a messenger from the depths of space.

Cosmic debris has been raining down on Earth ever since our planet formed. As well as tens of thousands of tonnes of dust, many thousands of pebbles and larger rocks are estimated to land on Earth every year, mostly fragments of the ancient asteroids that jostle and smash in a belt between Mars and Jupiter. Only a tiny fraction of falls are witnessed, and most of the material is lost in forests or oceans or worn to dust. But in some remote areas, such as the Antarctic icefields, these visitors can be preserved for thousands of years. In the 1970s the US space agency,

NASA, set up a mission to find them. In 1978 Score answered a job ad from the new Antarctic Meteorite Laboratory at NASA's Johnson Space Center in Houston, Texas, and six years later she made her first trip to Antarctica.

By the time Score made her find three days into the mission, she and the team of five other meteorite hunters working with her had picked up over a hundred meteorites between them. Straight away, though, she thought this one was special. It was large – grapefruit size – and it stood out as vivid green against the featureless blue ice. Back in Houston, it was Score's job to assign numbers to the team's finds, using the prefix ALH84 (for the location, Allan Hills, and the expedition year). She wanted her green mystery rock checked out first, so she put it top of the list. But, once unpacked, she was disappointed to find that it looked ordinary after all, like a piece of grey cement. The colour must have been caused by her tinted snow goggles, she figured, or a trick of the light. ALH84001 was later classified as an unremarkable chunk of asteroid and consigned to the laboratory stores.

It was another decade before this meteorite became a global sensation and household name. NASA researchers eventually realised that it was, in fact, a chunk of early Mars, far older than any other planetary rock ever discovered and containing a secret so explosive that it triggered a heartfelt speech from the US president himself. The discovery caused a worldwide media storm, sparked conspiracy theories and science-fiction plots, shifted NASA's future course and founded a new scientific field, astrobiology: the study of extraterrestrial life. Scientists are still arguing about what exactly is inside the rock that Score found on that Antarctic summer's day, but our understanding of life – on Earth and in the wider cosmos – would never be the same.

*

Are we alone in the universe? It's one of the biggest questions that humans have ever asked, deceptively simple to pose but for which either possibility – that we're the only vital spark in the entire barren eternity of existence or that we're just one bloom in a far bigger, cosmic web of life – is almost too epic to comprehend. It feeds into other enquiries, too, about who we are and what our existence means: What is life? Is humanity special? Why are we here at all?

The answers people give have ebbed and flowed through history, in step with their philosophical and religious beliefs. Ancient civilisations saw gods, spirits and souls in the sky. But the idea of physical aliens inhabiting other planets or solar systems goes back surprisingly far, too, at least to classical Greece. Followers of the 'atomist' school of philosophy, who believed reality is made up of tiny indivisible particles, argued that there are infinite atoms in the universe and

The first great outbreak of 'Mars fever' in the late 19th century was partly caused by a translation error. It all began in Milan, when the director of the Brera Observatory, Giovanni Schiaparelli, trained one of the latest telescopes on Mars and noticed he could make out details on the planet's surface: straight, dark grooves that he described as 'channels' and patiently reproduced in a series of maps. News of his discovery travelled around the world, but with the Italian *canali* translated as canals rather than channels, suggesting an artificial origin and giving rise to the theory that they were the work of intelligent beings – a theory that Schiaparelli himself was unwilling either to rule out or confirm. A particular proponent of the idea was the American astronomer Percival Lowell, who suggested that Mars was a dying planet whose civilisation was attempting to prevent desertification with the construction of gigantic waterways. Mars mania captured the public imagination (the scientist Nikola Tesla, for instance, claimed to have intercepted radio communications from the planet) and continued to persist even after the scientific community had rejected not just any extraterrestrial origin but also the very existence of canals on Mars – which they realised were the result of optical illusions created by the telescopes used at the time – largely thanks to the explosion in science-fiction writing. From Camille Flammarion's *Urania* to Ray Bradbury's *The Martian Chronicles* and Kim Stanley Robinson's more recent Mars trilogy, Schiaparelli's canals remain an integral part of the planet's fantasy landscape to this day.

therefore infinite worlds. The cosmologies of Plato and Aristotle that came to dominate Western thought had no room for other worlds or alien life: Plato argued that a unique creator implies a unique creation; Aristotle insisted that all elements find their natural place around a single centre, the Earth. Their Earth-centred teachings were later enforced as law by the Christian church, and for centuries, at least in the West, it was forbidden even to speculate about the existence of life elsewhere. It was the emergence of modern astronomy that opened the door. When Copernicus presented his heliocentric model for the solar system, says science historian Michael Crowe, he 'changed our Earth into a planet and transformed stars into other suns'. If ours was just one solar system among many, why shouldn't there be life on the others, too?

Enthusiasm for aliens reached a peak during the Enlightenment. As the huge extent of the cosmos became clearer, it seemed unimaginable to many that other life – even intelligent life – wouldn't exist somewhere. Prominent figures such as Immanuel Kant and Benjamin Franklin were in favour; legal scholar Montesquieu speculated about alien laws. In 1752 Voltaire satirised humanity's supposed intelligence in a novella that featured an extraterrestrial with more than a thousand senses. Much of the debate swirled around a central question: did divine creation extend throughout the whole universe (with humanity just part of a vastly larger plan), or did God create the cosmos purely to put us in it? Thomas Paine insisted the former when he rejected Christianity for deism in *The Age of Reason*. Opposing figures included the philosopher and scientist William Whewell, who argued in 1853 that the universe being so perfect for our needs proves that God designed it

'The crucial question regarding life elsewhere now came down not to God but to statistics: would life arise commonly given the opportunity, or is Earth an extraordinary fluke? The balance of evidence seemed to suggest the latter.'

just for us. Charles Darwin was scornful, privately describing Whewell's judgement that the solar system is adapted to us and not the other way around as an 'instance of arrogance!!'

For a while it seemed (to enthusiasts at least) that improved telescopes were bringing these aliens within reach. As far back as the 1770s William Herschel, pioneer of stellar astronomy, thought he saw forests and circular buildings on the moon. His son John, also a renowned astronomer, believed that life abounds in the solar system. In the 1860s he suggested 'huge, phosphorescent fishes' as an explanation for a report of giant leaf-shaped objects on the sun's surface. At the end of the 19th century US businessman Percival Lowell built an observatory in Arizona (at which Pluto was later discovered) and used it to detail what he claimed were networks of artificial waterways on the surface of Mars.

Then the wave of optimism crashed. The alien sightings didn't stand up to scrutiny, and by the mid-20th century scientists using improved methods of spectroscopy and infrared astronomy realised that the moon was dead and conditions on Mars were far harsher than thought. The possibility of Martian plants (perhaps some kind of lichen) was still widely accepted. But images from probes sent to Mars in the 1960s showed a bleak, barren world, and

hopes were finally dashed when NASA's Viking landers touched down in 1976. Ahead of the mission, the astronomer and science-populariser Carl Sagan excited the public with the idea of Martians perhaps as big as polar bears. 'The possibility of life, even large forms of life, is by no means out of the question,' he said. But no creatures wandered past the landers' cameras. Their instruments revealed that the atmosphere was vanishingly thin, and, with no protective magnetic field, the planet's surface was battered by deadly levels of radiation from the sun.

What's more, despite apparent positive results from experiments looking for signs of alien metabolism, other tests showed no detectable organics – molecules with a carbon backbone – in the soil. Since life on Earth is based on organic building blocks, it seemed the anomalous positive results must have a purely chemical explanation. Although not all of the researchers involved in the experiments agreed, NASA's conclusion was clear: there was no evidence for life on Mars.

Mars was seen as a test case, the most similar place in the solar system that we knew to Earth: if organisms existed anywhere else, it would surely be here. So the negative results severely knocked not just the chances of Martian life but life in the universe in general. 'Since Mars offered by far the most promising habitat for extraterrestrial life in the solar system,' wrote Norman Horowitz, who ran one of the Viking experiments, in 1986, 'it is now virtually certain that the earth is the only life-bearing planet in our region of the galaxy.'

Meanwhile biologists studying Earth life were realising that the steps required for life to start seemed staggeringly unlikely and complex. Since the structure of life's inheritable material, DNA, was worked out in the 1950s, they had been uncovering the intricate dance between DNA, RNA and proteins required to encode and transmit information from one generation to the next and the multilayered mechanisms by which these genetic instructions are carried out within a cell. For many, the chances of such a system arising spontaneously from scratch seemed so unimaginably small that it surely couldn't have happened more than once – a view supported by genetic studies showing that all known life on Earth is descended from a single ancestor.

The crucial question regarding life elsewhere now came down not to God but to statistics: would life arise commonly given the opportunity, or is Earth an extraordinary fluke? The balance of evidence seemed to suggest the latter. In the 1980s Francis Crick, co-discoverer of the structure of DNA, famously described the origin of life as 'almost a miracle' (in fact, Crick

PLANETARY DEFENCE

It is rare to see an asteroid strike Earth, but there are well-documented cases. In the so-called Tunguska event of 1908, an explosion in the atmosphere felled eighty million trees in Siberia. Also in Siberia, in 2013, thousands of people witnessed – and some of them filmed (you can find videos on YouTube) – another much smaller explosion over Chelyabinsk. In 1994 astronomers observed a comet breaking into several pieces and bombarding Jupiter with a shower of fragments that, had they fallen on Earth, would have wiped us out. (It is no coincidence that the films *Armageddon* and *Deep Impact* were released four years later.) Jupiter itself, thanks to its gravity, is our best protection against wandering space objects, but there is also a group of scientists calling for research funding in the field of planetary defence. The chances of a large object hitting Earth are minimal, but the fate of the dinosaurs teaches us that it is better to be prepared. NASA and the European Space Agency (ESA) have a programme to map Near-Earth Objects in order to identify the most dangerous in good time and, if needs be, destroy them or at least divert them. The suggested methods range from the simple (firing a projectile at them, as NASA has attempted with the 2021 launch of the DART probe designed to crash into an asteroid's moon) to the futuristic (destroying the asteroid with a laser). But there are also more Hollywood-style solutions (a good old nuclear warhead) and other more elegant schemes (painting one side of the asteroid black so that it absorbs more heat, which ought to alter its trajectory just enough to miss us and fly on by).

suggested that life didn't begin on Earth at all and was intentionally placed here by aliens, a view that was rejected by his peers); geologist Euan Nisbet summed up the prevailing view of life as 'an extraordinary accident on an extremely special planet'; Nobel-winning biochemist Jacques Monod pronounced that 'Man at last knows that he is alone'. The search for extraterrestrial life might engage science-fiction fans and conspiracy theorists but was no longer a serious topic of scientific enquiry. As cosmologist and bestselling author Paul Davies puts it: 'One might as well have professed an interest in looking for fairies.'

But then ALH84001 emerged from the stores.

*

David Mittlefehldt never intended to join the hunt for aliens. A geologist working at NASA's Johnson Space Center, he was interested in asteroids and what they can tell us about conditions in the early solar system. In 1988 he asked Score's lab to give him some pieces of asteroid to study, and one of the samples they sent him was from ALH84001.

Mittlefehldt probed the composition of the rock samples by firing beams of electrons at them, which causes different elements to give off characteristic patterns of radiation. The results for ALH84001 confused him at first, because they didn't match the other asteroid samples he was studying. It was several years before he finally accepted the reason: this rock had been misidentified. In fact, its composition matched a tiny group of meteorites known as SNCs, named after three stones that fell in Shergotty, India, El Nakhla, Egypt, and Chassigny, France. Other researchers had shown just a few years earlier that the mixture of trapped gases they contain perfectly matches the Martian atmosphere as measured by the Viking landers. These rocks came not from asteroids but from Mars. Just nine meteorites were known from this select group. In October 1993 Mittlefehldt announced that he had found a tenth.

The news spread fast – Score was thrilled – and researchers around the world rushed to study pieces of her favourite rock. They soon realised that even among SNCs ALH84001 was special, having formed from volcanic lava over four billion years ago. That made it incredibly ancient, almost as old as the solar system itself, more than three times as old as the next-oldest Mars meteorite and significantly pre-dating any rock known on Earth. Studies of molecular signatures in the rock showed that it was ejected into space by an impact on Mars sixteen million years ago and drifted through the solar system until thirteen thousand years ago, when it was captured by Earth's gravity and landed in Antarctica. There it became imprisoned, deep in the ice, until the region's relentless winds exposed it again for Score to find.

And there was something else curious. The rock is full of tiny cracks, from an earlier impact that occurred when it was still in place on early Mars, and the surfaces of these fractures are dotted with tiny flattened grains. Sometimes described as 'moons' or 'globs', they are only just visible to the naked eye, but under the microscope they appear as golden circles with black-and-white rims. They're made of carbonate, a type of mineral containing carbon and oxygen. On Earth, rocks containing carbonate (such as limestone or chalk) most commonly form in water – for example, when shells and skeletons from marine creatures accumulate and fossilise. Carbonates aren't usually found in meteorites, so Mittlefehldt asked Chris Romanek, an expert in

Its theoretical existence was put forward by the astronomers Mike Brown and Konstantin Batygin, and if proven would be a momentous discovery: a new planet in our solar system, between six and ten times larger than Earth and provisionally named Planet Nine or Planet X. Is it possible that we would not have noticed it before? Yes, because the planet is thought to be four hundred times further away from the sun than we are and almost ten times as distant as Pluto. Brown has long studied the peripheral region of our solar system known as the Kuiper belt and already made his name identifying numerous objects beyond Neptune, some of them large. He is also known as 'the man who killed Pluto' for his role in having it downgraded to a dwarf planet after discovering another dwarf planet, Eris, in 2005, which has a mass a third greater than Pluto's. Other scientists have noted that a series of planetoids in the region follow abnormal or incomprehensible orbits: the explanation put forward by Brown and Batygin is that the anomaly is down to the gravitational disturbances caused by a gigantic planet, perhaps a 'super-Earth'. The same technique led to the discovery of Neptune (1846) and Pluto (1930), but, when it comes to planets, the scientific community is of the opinion that seeing is believing. So while the pair spend hours at their telescopes in the hope of capturing the image that would make them famous, the debate has become a hot topic, with some convinced there is more than one planet still to discover, others believing that the orbits are not disturbed by just one large object but by myriad smaller ones and even those who suggest that Planet X was already known in the Middle Ages.

carbonates working in the same building, to look at them.

Working with NASA geochemist Everett Gibson, who had helped probe the moon rocks brought back by Apollo astronauts and was now studying Martian meteorites, Romanek blasted the carbonates with lasers to analyse the carbon and oxygen they contained. The results suggested that the carbonates had formed on Mars (rather than being contamination from Antarctica) and were deposited inside the rock after carbon dioxide, dissolved in liquid water, had flowed into the cracks. Today's Mars has an average temperature of −60 degrees Celsius, too cold for liquid water. But ALH84001 pointed to a far more hospitable past.

That wasn't all. The pair also saw microscopic shapes near the carbonate grains: worms and sausages that looked just like Earth bacteria, except much smaller. It was the beginning of a crazy idea. Could water flowing into the rock have carried not just carbonates but tiny Martian microbes? In September 1994 Gibson and Romanek showed their images to David McKay, a senior NASA scientist who a few decades earlier had taught geology to the Apollo astronauts. They agreed to pursue the idea, in secret, along with one more person: a few days later, McKay approached Kathie Thomas-Keprta, a NASA chemist and specialist in electron microscopy. 'I kind of thought he was crazy,' she said later, but she reluctantly agreed to help investigate the rock's strange features. 'I thought I would join the group and straighten them out.'

It wasn't long before she, too, was hooked.

*

Across the Atlantic, at an observatory nestled among the lavender fields of

south-east France, a young astronomer named Didier Queloz was wondering if he was crazy. He had spent the summer of 1994 analysing results from a new instrument, a spectrograph that could analyse the movements of distant stars more accurately than ever before in the hope of spotting distant planets. It had been suggested that a planet orbiting a star would exert a gravitational tug, causing the host star to wobble slightly, which might be detectable. But Queloz's results made no sense at all. In particular, one star, 51 Pegasi, wasn't behaving. Its speed was unstable, cycling in a pattern that repeated every four days. At first Queloz, who was studying for his PhD at the University of Geneva, thought it was a bug in the software he had designed. Panicking, and too ashamed to tell his supervisor Michel Mayor that the instrument wasn't working properly, he spent the autumn testing everything he could think of and repeatedly observing the star. But the wobble refused to disappear.

Eventually, he allowed himself to consider whether the anomaly might be a gravitational pull after all. He calculated that to explain the wobble would take a giant half the mass of Jupiter and around 150 times the mass of Earth, yet so close to its star that instead of taking a year to orbit, like Earth, it completes the trip in just four days. Such a star-hugging monster was unlike anything in our solar system, against all accepted theories of planet formation, and it contradicted respected astronomers who had searched for planets and concluded there was nothing out there with a period of less than ten years. But Queloz couldn't see any other explanation. He sent a fax to Mayor in Hawaii: 'I think I found a planet.'

Mayor returned to Europe in March 1995, but by this time 51 Pegasi had dipped below the horizon. So the pair calculated predictions for the orbit of the putative planet, ready for when the star reappeared in the night sky in July. When the time came, they travelled with their families to Haut-Provence to compare the behaviour of the rogue star against their model. The first night its speed matched their predictions perfectly. And on the second night, and the third, and the fourth. On the fifth night, recalls Queloz, 'we said yes, it's a planet'. Mayor later described the realisation as 'a spiritual moment'. They were the first humans in history to know that other worlds really do exist beyond our solar system. Nearly fifty light years away, in the constellation Pegasus, a giant gaseous planet was circling a star similar to our sun. They named it 51 Pegasi b, and celebrated with local cake and sparkling wine.

Mayor and Queloz announced the news at a conference in Florence in October 1995. The astronomers had realised that 51 Pegasi b, as the first extrasolar planet, was a milestone for astrophysics, plus its orbit was so strange that theories of planet formation would have to be rewritten. But the public reaction took them completely by surprise, with front-page headlines around the world and so many calls and interview requests, says Queloz, that for the next six months it was impossible to work. 'What we completely missed,' he told me, 'was the connection between planet and life.'

51 Pegasi b itself is a searingly hot, inhospitable world locked close to its sun and glowing at 1,000 degrees Celsius. The chance for life there appears to be approximately zero. But the planet's mere existence implies something far greater: if there was one planet out there, there must be more. 'Other worlds are no longer the stuff of dreams and philosophic musings,' commented veteran space reporter John Noble Wilford in *The New York Times*.

'In thousands of years of speculation about alien life, we'd never had a scrap of actual evidence to hint that we are anything but utterly alone. Now, at last, that had changed.'

'They are out there, beckoning, with the potential to change forever humanity's perspective on its place in the universe.' In thousands of years of speculation about alien life, we'd never had a scrap of actual evidence to hint that we are anything but utterly alone. Now, at last, that had changed.

*

Back in Houston, Thomas-Keprta's plan at first was to save McKay and the others from embarrassment, to prove to them that there was no evidence of Martian life in ALH84001. As she scanned the meteorite's microscopic terrain, with its orange moons embedded in the silver-grey rock, she became intrigued by tiny black grains in the rims of the carbonate globules, just nanometres across. She found that they are magnetic crystals, made of magnetite (iron oxide) and pyrrhotite (iron sulphide), just like the tiny compasses produced by magnetotactic bacteria on Earth. There are non-biological ways to make these minerals, but this generally takes extreme conditions – high temperature and pH – so it was hard to explain how they ended up in carbonates deposited at mild temperatures. Unless these crystals were made by bacteria, too.

Thomas-Keprta sent two flecks of ALH84001 to chemist Richard Zare at Stanford University. He had a powerful instrument called a laser mass spectrometer that could identify even trace amounts of chemical molecules by vaporising them. Zare and his colleagues found what Viking had failed to: organics from Mars. Complex organic molecules called PAHs were concentrated within the carbonate moons. PAHs can form non-biologically – they're in everything from car exhaust to interstellar gas clouds – but on Earth you also find them wherever life has been, for example in petroleum or coal. The molecules that Zare and his colleagues found in ALH84001 were exactly what you would expect to see when bacterial cells decay.

For the team, the coming together of so many potential fingerprints of life had to be more than a coincidence. In early 1996, with excitement combined with 'gut-fluttering dread', they submitted their paper to *Science*. After weeks of scrutiny by a panel of reviewers (including an ageing Carl Sagan) it was accepted for publication, setting in train a sequence of events that reached the country's highest levels of power. In July the team was called into the office of NASA chief Dan Goldin, who grilled them for hours, then Goldin, in turn, was summoned to the White House to brief President Bill Clinton and Vice-President Al Gore.

At a press conference on Wednesday 7 August, President Clinton addressed the world live from the White House Rose Garden. 'Today, rock 84001 speaks to us across all those billions of years and millions of miles,' he said. 'It speaks of the possibility of life.' If confirmed, he added, the implications 'are as far-reaching and awe-inspiring as can be imagined'. Then the TV networks switched to the packed main auditorium at NASA headquarters. One by one, the team – including McKay, Gibson, Zare and Thomas-Keprta – described their lines of evidence: the carbonates, magnetite crystals and organics. None was conclusive on its own, but taken together, the team argued, they were evidence for primitive life on early Mars. They played an animation showing worm-like microbes on Mars swimming in water and becoming trapped in the build-up of carbonate deposits inside cracks in the rock, before the rock was sent spinning into space and eventually landed in Antarctica. Finally, to an audible gasp from the audience, they showed their images of the putative 'fossils' themselves.

News crews lined up at the meteorite lab in Houston for a glimpse of the original rock, while the scale of coverage in the world's newspapers and magazines eclipsed even that of the first moon landing.

Not all the coverage was positive. Zare's lab had to temporarily take down its website and contact details after religious fundamentalists complained that its research contradicted the Bible. And many scientists, too, were gearing up for a furious response.

*

Over the coming months other researchers vehemently attacked not just the ALH84001 paper but its authors. The critics challenged every line of evidence, arguing that the worm shapes were created in the lab; that the carbonates and organics were simply contamination from Antarctica; or that the carbonates, even if Martian, had formed at temperatures much too high for life.

The team's observations stood up to scrutiny, and there is now consensus that the features they reported did form on Mars and that the carbonates were deposited in a watery environment at temperatures around 25–30 degrees Celsius. The debate shifted instead to the interpretation of those results. Whereas McKay and the others argued that adding several 'maybes' together strengthened their case, others dismissed the carbonates, organics and worm-like shapes as explainable in other ways.

The NASA team has been studying other Martian meteorites, including Nakhla and Shergotty, and found suggestive evidence there, too. In 2014 the researchers reported complex organic matter stuffed into veins and cracks of a Mars rock known as Yamato 000593, as well as tiny tubules that look just like tunnels found in some Earth rocks (ancient and modern) thought to be etched by microbes hunting for nutrients. McKay passed away in 2013, but Gibson and Thomas-Keprta are still working on the project. 'We continue to support our original hypothesis,' they told me in June 2019. But the debate has reached a frustrating stalemate. Although the NASA team is as convinced as ever that life is the most plausible explanation, critics still insist that they have failed to prove their case. What everyone does agree on, though, is that the team's work, and the surge of interest that followed, helped to transform the search for alien life.

In the early 1990s NASA was struggling to find a purpose. With the collapse of the Soviet Union the space race was over, and the agency's huge-budget missions were suffering from delays, cost overruns and major failures, including the loss of Space Shuttle *Challenger* in 1986 and the $1 billion *Mars Observer* orbiter in 1993. The

Looks Lovely, But I Wouldn't Live There

EMANUELE MENIETTI
Translated by Alan Thawley

The moon

AVERAGE DISTANCE FROM THE EARTH: ★ ★ ★
Around 385,000 kilometres from Earth, practically around the corner in astronomical terms.

CLIMATE: ★ ☆ ☆
Either too hot or too cold – it ranges between 127°C in full sun to −173°C in the shadows.

HABITABILITY: ☆ ☆ ☆
The lack of breathable atmosphere and the absence of liquid water mean life, at least as we know it, is impossible. Thanks to the Apollo missions, however, the moon is the only celestial body apart from Earth known beyond any doubt to have hosted life, even if only for a short time.

The moon is a small, rocky, inhospitable world. With a diameter of 3,474 kilometres and a mass one-eightieth of the Earth's, it is the fifth largest natural satellite out of almost two hundred in our solar system. Its atmosphere is extremely tenuous, with a density comparable to the outermost layers of the Earth's. NASA's planned Artemis programme will take the first female astronaut to the moon. China, meanwhile, attempted to grow cotton there. Despite being insulated from the lunar environment, the plants died after a couple of days.

Mercury

AVERAGE DISTANCE FROM THE EARTH: ★ ★ ★
At 155 million kilometres from the Earth, it is the closest planet to the sun.

CLIMATE: ★ ☆ ☆
Surface temperature ranges between 430°C and −180°C.

HABITABILITY: ★ ☆ ☆
Very primitive micro-organisms may have lived underground on Mercury, but not everyone is convinced by this.

If you are planning a move to space, you can cross Mercury off your list. While it may be a rocky planet like Earth, its proximity to the sun and the very thin atmosphere make it inhospitable, with temperatures that fluctuate by hundreds of degrees. On the up side, one solar day on Mercury lasts for 176 Earth days. The planet measures just 4,879 kilometres across, less than half the Earth's diameter, and its irregular surface is dotted with craters reminiscent of the moon. It has been geologically inactive for billions of years but is nonetheless interesting: the joint European and Japanese *BepiColombo* space mission will soon be visiting to carry out studies and discover more about planets that orbit close to their host stars.

Venus

**AVERAGE DISTANCE FROM
THE EARTH:** ★★☆
At 170 million kilometres from Earth,
it is the most luminous celestial body
visible from Earth after the moon.

CLIMATE: ★★☆
With an average temperature of
464°C, it is hotter than Mercury,
despite being further from the sun.

HABITABILITY: ★☆☆
There is a theory that its extremely dense
atmosphere may be home to life forms of
some kind, but there is a hot debate around
the first clues that might support this.

Its dimensions and mass are similar to the
Earth's, but its atmosphere consists mainly
of carbon dioxide, which contributes to a
greenhouse effect unlike any other in the
solar system and makes it the hottest of our
neighbouring planets. Direct observation
of its surface is extremely difficult because
of the very dense clouds consisting mainly
of sulphuric acid. At an altitude of sixty
thousand metres, atmospheric pressure and
temperatures are more moderate, leading
to theories that life could be supported in
that zone. In 2020 a group of researchers
said they had found promising clues, but
extraordinary claims require extraordinary
proof, which no one has yet produced.

Mars

**AVERAGE DISTANCE FROM
THE EARTH:** ★★☆
Around 254 million kilometres, but every
two years it is closer to Earth and could be
reached in six to eight months at that time.

CLIMATE: ★★☆
It can get seriously cold (−143°C),
but in 'summer' on the equator air
temperatures reach a balmy 35°C.

HABITABILITY: ★★☆
Conditions were once much more favourable
to life, but Mars is now an arid planet, with
reserves of frozen water at the poles. There
could still be liquid water underground,
which would make life more probable.

While we do not yet know whether microbial
Martians exist (or ever have) what we can be
sure of is that the planet has a growing robot
population. We continue to send them out to
explore the rocky planet, which is similar to Earth
in some respects. It would give Mount Everest
an inferiority complex, however: Olympus Mons,
the highest mountain on Mars (and in the solar
system, excluding asteroids), stands twenty-
one thousand metres tall. Mars is also in a
league of its own when it comes to sandstorms;
whipped up by winds of up to 160 km/h, they
sometimes envelop the entire planet. While we
wait for the opportunity to explore in person,
our robots continue to do an excellent job.

Jupiter

**AVERAGE DISTANCE FROM
THE EARTH:** ★☆☆
We are 787 million kilometres away
from the fifth planet from the sun.
CLIMATE: ★☆☆
Its dense clouds surround an icy world
with temperatures of −145°C, but things
get a lot hotter towards the centre of
the planet, where the temperature is
estimated to exceed 20,000°C.
HABITABILITY: ☆☆☆
Jupiter is one of the last places to
look for life as we know it.

It would take eleven planets like ours side by
side to match its diameter, and three hundred
to equal its mass. The largest planet in the solar
system is not solid, however: it is a ball of gas
consisting mainly of helium and hydrogen that
formed by swallowing up the remnants of the
matter that made up the sun. Some, indeed,
regard it as a failed star. Jupiter looks like a huge
marble made up of coloured layers: these are
the dense cloud formations that cover the outer
surface of its atmosphere. The striped effects
are caused by strong winds and giant storms.
The storm that gave rise to the famous Great
Red Spot has lasted at least three centuries and
is the largest in the entire solar system: it is big
enough to contain the Earth almost three times
over. On Jupiter, everything is on a gigantic scale.

Enceladus

**AVERAGE DISTANCE FROM
THE EARTH:** ★☆☆
At 1.4 billion kilometres away, it is pretty
much on the outskirts of the solar system.
CLIMATE: ★☆☆
At midday it hits an average
temperature of −198°C, and the surface
is covered in a thick layer of ice.
HABITABILITY: ★★★
The presence of an ocean and
organic molecules make it an ideal
place to support life forms.

If you like to bet, put some money on
Enceladus in the search for extraterrestrial
life. Saturn's sixth largest moon resembles
a desolate ball of snow, with a respectable
diameter of five hundred kilometres.
There is a surprise concealed beneath its
35-kilometre-thick shell of ice, however:
an ocean of water with a rocky bed that,
according to researchers, could play host
to the complex chemical reactions that are
essential for life. Conditions on Enceladus
are turbulent, and it periodically emits
jets of gas that penetrate the icy shell:
from the analysis it seems they contain
traces of hydrogen, another indicator
of potential biological activity. Various
missions are being planned to go and
check it out using probes and robots.

Kepler-452b

**AVERAGE DISTANCE FROM
THE EARTH:** ☆☆☆
At around 550 parsecs away from
our solar system, it is completely out
of reach: light from the planet takes
around 1,800 years to reach us.

CLIMATE: ★★★
If the Earth had no atmosphere and was
only heated by the sun, its planetary
equilibrium temperature would only
be a fraction higher than the figure of
−8°C for this extremely distant world.

HABITABILITY: ★☆☆
It is located within the habitable zone, in
other words at the right distance from its
host star to be neither too hot nor too cold,
a good starting point for supporting life.

Kepler-452b is an exoplanet, meaning
that it is located outside our solar system.
NASA has described it as a 'cousin of the
Earth', not because of any kinship but more
down to the theory that it shares similar
characteristics. It is likely to be rocky, with
a mass five times greater than that of the
Earth, and orbits a star similar to the sun.
Kepler-452b is a couple of billion years older
than our own planet, however, and may lack
an atmosphere. And it wouldn't be easy to
get there: with our fastest probe to date, it
would take 26 million years to make the trip.

Proxima Centauri b

**AVERAGE DISTANCE FROM
THE EARTH:** ★☆☆
It is the closest planet outside the solar
system to be discovered so far, but it's
still almost forty trillion kilometres away.

CLIMATE: ★☆☆
Its equilibrium temperature is
estimated to be −39°C, and it may
have an atmosphere of hydrogen and
helium similar to the planet Neptune,
which is still extremely cold.

HABITABILITY: ★☆☆
It is located in the habitable zone but
is probably lashed by powerful stellar
winds, meaning that life could only
be possible beneath its surface.

Discovered in 2016, Proxima Centauri b
has a mass estimated to be 1.3 times
greater than Earth's. It orbits Proxima
Centauri, a red dwarf, which has a much
lower luminosity than the sun. There
is an unconfirmed hypothesis that it is
rocky; it is also unclear whether it has an
atmosphere that would make the presence
of life more likely. There are plans to build
spacecraft equipped with solar sails, which
could use radiation pressure to move at
20 per cent of the speed of light. If this
comes off, Proxima Centauri b could be
reached by a probe in a mere twenty years.

SO NEAR AND YET SO FAR

As of 1 April 2022, there are 4,984 confirmed exoplanets in 3,673 planetary systems, with 815 systems having more than one planet. A few thousand more are categorised as candidates yet to be confirmed. It is very difficult (although not impossible) to observe exoplanets directly, so the overwhelming majority of those that we know of have been observed indirectly through the effects that they induce in their host stars, in particular thanks to the transit and radial velocity methods.

TRANSIT METHOD

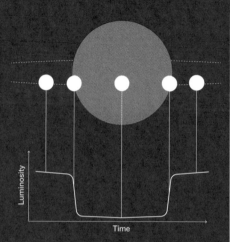

RADIAL VELOCITY METHOD

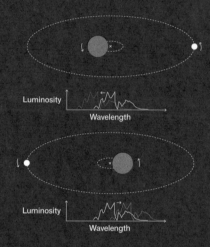

The transit technique, or photometric method, has been responsible for the discovery of around three-quarters of all known exoplanets, and the first detections date back to 1999. When a planet passes – transits – in front of its star it causes a measurable reduction in the star's luminosity. The transit is repeated at intervals governed by the time it takes the exoplanet to complete its orbit around the star. For example, an observer of our solar system would have to wait a year to see a repetition of the Earth's transit in front of the sun. An exoplanet with the same mass as Jupiter causes a reduction in the luminosity of its host star of around 1 per cent. The photometric method is therefore most suited to the discovery of large planets and – unless you want to spend whole months waiting for the following transits – those with short orbital periods.

This was the method used by Queloz and Mayor to discover the first exoplanet, 51 Pegasi b. Rather than rotating exactly around its own centre, a star with a planet orbiting it rotates around the combined centre of gravity of the star and the planet (shown with an 'x' in our simplified diagram). This means that when observed from our viewpoint, the star is not immobile and instead 'oscillates' because of its circular motion, coming nearer and further away from us. Viewed through special instruments known as spectrographs, the light of a star coming towards us will look more blue (blueshift), whereas when it is going away from us the light will look more red (redshift). This change is a relative Doppler effect that enables us to measure the star's speed of rotation around the shared centre of gravity and therefore to deduce the presence of a planet and also to estimate its minimum mass, its orbit and even its chemical composition – but at this point things start to get much more complicated.

SOURCE: EUROPEAN SPACE AGENCY

White House was cutting billions from NASA funding; indeed, much of the vitriol directed towards ALH84001 was because planetary scientists feared it would strengthen the argument that taxpayers' money was better spent elsewhere. 'We're at the bottom of the pecking order in NASA's budget,' meteorite expert Allan Treiman told *Newsweek* in 1997. 'People are concerned that if this turns out to be as stupid as cold fusion we'll be out on the street.'

In fact, the opposite happened. Dan Goldin was already streamlining the agency by slashing bureaucracy, cutting jobs and pushing small, innovative missions with the motto 'faster, better, cheaper'. But that wasn't all. He envisioned a new – scientific – mission for NASA: to answer big questions about the universe and our place within it. He had to convince the politicians, though, and ALH84001 came at just the right time. When Clinton addressed the world to announce news of the meteorite, he described the research as a 'vindication of America's space program and our continuing support for it, even in these tough financial times'. He promised that NASA would 'put its full intellectual power and technological prowess behind the search for further evidence of life on Mars'.

Clinton halted the decline in NASA's funding, and the agency shifted resources towards planetary exploration. Mars missions were revived with a new generation of spacecraft designed – for the first time since Viking – in seeking biomarkers and habitats relevant to life. In 1998 NASA founded its Astrobiology Institute, and a new field was born that now extends far beyond this one agency. The multipronged research into ALH84001, says Gibson, provided 'the guiding idea'. Rather than considering 'aliens' in isolation, scientists from a broad range of disciplines – understanding how planets form, studying past life on Earth, detecting organics in clouds of interstellar gas – would now work together with one overarching aim: to understand the cosmos as it relates to life.

*

The research that followed has reshaped our view of the cosmos within just a few decades. Take the hunt for planets outside our solar system, or 'exoplanets'. Within weeks of the announcement of 51 Pegasus b in 1995, researchers started finding more. But the field hit warp speed when NASA launched the Kepler space telescope, which scanned stars for a slight dimming of their light caused by a planet passing in front. It became the most prolific planet hunter ever, taking the count of known exoplanets to more than four thousand before it was decommissioned in 2018 (with thousands more candidates waiting to be confirmed). Kepler showed that planets are not only very common in our galaxy but breathtakingly diverse, from hot Jupiter-like 51 Pegasus b to mini-Neptunes perhaps covered by planet-wide oceans. Many are tidally locked 'eyeball worlds', with endless night on one side, searing day on the other and a knife-edge of permanent twilight in between.

Researchers are now using spectroscopy to probe planets' chemical composition, and here, too, they are finding an array of exotic specimens far beyond anything that exists in our own small corner of the galaxy. Examples include 55 Cancri e, a dense scorching volcano world that orbits its star every eighteen hours; HAT-P-7b, a gas giant with clouds of vaporised corundum, the mineral that makes rubies and sapphires; and Kepler-7b, as light as polystyrene.

Somewhere hot, somewhere cold, deprived of water, in a vacuum ... they stay wherever you put them without making too much fuss. The real stars of astrobiology are little creatures just a few tenths of a millimetre long. Stocky, with eight retractable feet, and affectionately known as 'water bears', tardigrades are eukaryotic invertebrates able to survive in extreme conditions. They have been observed on all continents, including Antarctica, and at every altitude, from the depths of the ocean to six thousand metres above sea level in the Himalayas. What makes them special are their defensive mechanisms in response to hostile environments: some are able to expel all the water from their bodies and 'turn off' their metabolism. In this state of hibernation, or cryptobiosis, they have survived after being kept for months at −200 degrees Celsius (others, on the other hand, have lived through several minutes at a temperature of 150 degrees Celsius) and can even remain for up to a hundred years in a state of total dehydration; as soon as they come back into contact with water, all their vital functions resume. Scientists have subjected them to every kind of test: extreme pressures, they emerge unscathed; bombarded with UV rays, fit as a fiddle; deprived of oxygen, a walk in the park. In one experiment they were left for ten days in the vacuum of space at the mercy of solar radiation; on their return to Earth most of them were still alive and even able to reproduce. A 2017 study suggested that tardigrades would survive the type of astronomical events – asteroid strikes, gamma-ray bursts, supernova explosions – that could wipe out human life. Blessed be the tardigrades, for they shall inherit the Earth.

Kepler-16b has a double sunset like *Star Wars'* Tatooine, while HD189733b sports a vibrant blue atmosphere and silicon clouds that rain glass.

Central to the effort is the hunt for planets that might harbour life, which (in the absence of a better description) astronomers define as planets similar in size to Earth and at the right distance from their star for liquid water to exist. These planets seem to be abundant, too. A 2013 analysis of Kepler data concluded that around a fifth of all stars in the Milky Way are orbited by at least one such world. In 2016 a rocky planet was discovered in the habitable zone of our nearest star, Proxima Centauri; the next year astronomers found a system with seven Earth-size planets circling the same sun (three in its habitable zone). In 2019 water vapour was detected in the atmosphere of a planet in the constellation Leo. Overall, astronomers estimate there could be around 8.8 billion potentially habitable Earth-size planets in our galaxy alone.

That's a huge shift in perspective, just as those commentators predicted after 51 Pegasus b was revealed in 1995. A few decades ago the idea of 'other worlds' was pure fiction, as it had been for millennia. Now we're confronted with a diversity of planets beyond anything we could have imagined – a universe with more planets than stars. Even if life is vanishingly unlikely to arise on any particular planet, we know that in our galaxy alone there are billions of chances for it to occur.

At the same time, biologists have been realising that life itself is far more flexible and tenacious than they ever thought. The discovery of thriving ecosystems around deep-sea hydrothermal vents came as a shock in 1977. But since the study of 'extremophiles' (organisms that favour extreme conditions) took off

'Biologists have been realising that life itself is far more flexible and tenacious than they ever thought. The discovery of thriving ecosystems around deep-sea hydrothermal vents came as a shock in 1977.'

in the 1990s, it has become clear that on Earth, at least, wherever there is even a hint of liquid water, there is life. 'What we previously thought of as insurmountable physical and chemical barriers to life,' biologists commented in *Nature* in 2001, 'we now see as yet another niche harbouring extremophiles.'

As recently as 2013 bacteria were found living in frigid, briny lakes beneath Antarctic ice and deep within the super-heated rocks of the Earth's crust. Entire ecosystems glean their energy not from sunlight, as once thought universal, but chemical energy from deep inside the planet. Bacteria can thrive in conditions of high acid or salt, extreme gravity, crushing pressures and harsh radiation; there are strains that can feed on uranium or breathe arsenic. Lichen grows in Martian conditions. Tiny but tough micro-animals called tardigrades have laid eggs in the near-vacuum of space. Each new discovery has expanded our notion of life, and, in turn, made it easier to imagine that it could have evolved elsewhere.

We're increasingly appreciating other potential habitats in the solar system, too, such as Europa and Enceladus (moons of Jupiter and Saturn respectively), where astronomers believe vast water oceans lie under kilometres of ice; or Venus, which could have been habitable billions of years ago before it

was scorched by a runaway greenhouse effect. Ideas about Mars itself, past and present, have also rebounded. In 1997 measurements made by NASA's orbiter *Mars Global Surveyor* showed that four billion years ago Mars had a planet-wide magnetic field that would have protected it from the damaging 'wind' of charged particles from the sun. There's now strong evidence from Martian rocks that this allowed the planet to retain a thicker atmosphere, rich in carbon dioxide, which kept the climate relatively warm, with liquid water in briny lakes, rivers and seas. In 2018 NASA's *Curiosity* rover sampled three-billion-year-old sedimentary rocks from the bottom of an ancient lake and found plenty of the organic building blocks necessary for life. It all fits eerily well with the story told by McKay and the others back in 1996, of magnetotactic bacteria that once lived in salty, carbonated water.

Today the magnetic field is gone, leaving Mars beaten by harsh radiation and with an atmosphere so thin that liquid water soon boils away. Yet data from Mars missions has consistently painted a more life-friendly picture than expected. Scientists have seen plenty of frozen water, in polar ice caps and underground, as well as dark streaks on equatorial hillsides during summer that look like flowing water. In 2018 the European Space Agency's *Mars*

Express orbiter used radar to show that, just like in Antarctica, there's a large lake of liquid water, twenty kilometres across, deep below the ice at Mars' south pole. Methane has also been repeatedly detected in the planet's atmosphere. Although this gas can be produced geologically, most terrestrial methane is made by organisms, including bacteria living in the Earth's crust, so some scientists believe it might signal the presence of similar microbes below Mars' surface.

Overall, the difficulties inherent in interpreting the ALH84001 and Viking results have forced scientists to think more broadly about how to know aliens when they see them. Just as planets can be staggeringly diverse, maybe living creatures can, too. The life we know is carbon based and reliant on liquid water as a solvent, but different chemistries have been proposed: perhaps life in the methane/ethane lakes of Saturn's moon Titan could use these hydrocarbons as a solvent instead of water. Perhaps different energy sources are possible: instead of sunlight or chemical energy as on Earth, maybe thermal energy or kinetic energy. With so much potential variety, should we look for the ability to evolve, or to metabolise, or to encode information? Or perhaps any definition we come up with will be too limiting, and we should just look for something that doesn't fit.

One speculative possibility is that beings on a planet orbiting a neutron star might harvest energy from fluctuating magnetic fields, basing their genetic code on chains of magnets instead of chemical DNA. Others have suggested that a 'shadow biosphere' of alien life, undetected by our conventional tests, might exist on Earth, 'like the realm of fairies and elves just beyond the hedgerow'. Whether or not we find such creatures, the very search is changing us by stretching our ideas about what life itself is.

*

Despite all the discoveries since 1996, we still don't have a single proven example of life elsewhere. For the first time in human history we can use scientific methods to investigate empirically what's out there in the universe, but we can't yet answer the big question – is there other life? – on Mars or anywhere else. What we have done very successfully, though, is break down the search into a series of smaller questions: Are other planets common? Is Mars habitable? Are organic molecules widespread? Can organisms survive in extreme conditions, even in space? And it's hard to ignore the fact that, so far at least, the answers keep coming back 'yes'. Over the past few centuries science has demoted our existence from God's special creation to a chance aberration in an otherwise empty universe. Score's Antarctic stone and Queloz's whirling planet help to point towards a different possibility: life as a common occurrence. Wherever we look, the evidence seems to support the case that life in the cosmos is not the exception but the rule. ✒

This article is an adapted extract from *The Human Cosmos* by Jo Marchant, published in 2020 by Dutton in the USA and Canongate in the UK.

The Eighth Continent: The Race to Develop the Moon

Only twelve men have ever set foot on the moon, the last in 1972. Fifty years later another space race is under way, but the rockets are no longer fuelled only by the rivalry between great powers: China, the USA and others – alongside dozens of private companies – are motivated now by the desire to make a profit as well as the advancement of science. The moon is back in fashion and is about to lose a little more of its mystery.

RIVKA GALCHEN

Top to bottom: Apollo 11 lunar module in lunar landing configuration; Apollo 15 command and service module in lunar orbit; S-IVB stage used by NASA; Soyuz spacecraft; Apollo 17 transposition/docking manoeuvres. Original images: NASA

RIVKA GALCHEN is a Canadian-American writer, whose work has received numerous awards, including the William Saroyan International Prize for Writing and the Rona Jaffe Foundation Writers' Award. Her first novel was released in 2006, and since then she published four more books of fiction and non-fiction, her most recent being her second novel, 2021's *Everyone Knows Your Mother Is a Witch* (Farrar, Straus and Giroux, USA / Fourth Estate, UK). She contributes to *The New Yorker*, which, in 2010, included her in a list of the twenty best writers under forty. She also regularly writes for a number of other journals, including *Harper's Magazine*, *The New York Times*, *The Believer* and the *London Review of Books*.

The staff of the photographic agency Prospekt created this series of images by digitally processing original photographs of the Apollo missions and ESA's Galileo System held in the NASA archives.

I n January 2019 the China National Space Administration landed a spacecraft on the far side of the moon, the side we can't see from Earth. *Chang'e-4* was named for a goddess in Chinese mythology who lives on the moon for reasons connected to her husband's problematic immortality drink. The story has many versions. In one, Chang'e has been banished to the moon for elixir theft and turned into an ugly toad. In another, she has saved humanity from a tyrannical emperor by stealing the drink. In many versions, she is a luminous beauty and has as a companion a pure-white rabbit.

Chang'e-4 is the first vehicle to alight on the far side of the moon. From that side, the moon blocks radio communication with Earth, which makes landing difficult, and the surface there is craggy and rough, with a mountain taller than anything on Earth. Older geologies are exposed, from which billions of years of history can be deduced. *Chang'e-4* landed in a nearly six-kilometre-deep hole that was formed when an ancient meteor crashed into the moon – one of the largest known impact craters in our solar system.

You may have watched the near-operatic progress of *Chang'e-4*'s graceful landing. Or the uncannily cute robotic amblings of the lander's companion, the *Yutu-2* rover, named for the moon goddess's white rabbit. You may have read that, aboard the lander, seeds germinated (cotton, rapeseed and potato; the Chinese are also trying to grow a flowering plant known as mouse-ear cress) and that the rover survived the fourteen-day lunar night, when temperatures drop to −168 degrees Celsius. *Chang'e-4* is a step in China's long-term plan to build a base on the moon, a goal towards which the country has rapidly been advancing since it first orbited the moon in 2007.

If you missed the Chinese mission, maybe it's because you were focused on the remarkably inexpensive spacecraft from SpaceIL, an Israeli nonprofit organisation, which crash-landed into the moon on 11 April 2019 soon after taking a selfie while hovering above the lunar surface. The crash was not the original plan, and just two days later SpaceIL announced its intention of going to the moon again. But maybe you weren't paying attention to SpaceIL either, because you were anticipating India's *Chandrayaan-2* moon lander, launched in July 2019. Perhaps you were distracted by the announcement, on 21 January 2019, the night of the super blood wolf moon, that the European Space Agency plans to mine lunar ice by 2025. Or by Vice-President Mike Pence's statement, in March 2019, that the United States intends 'to return American astronauts to the moon within the next five years'.

Half a century ago three men journeyed from a small Florida peninsula to a dry crater some 384,000 kilometres away called the Sea of Tranquillity. Hundreds of millions of people watched on black-and-white TVs as a man from Wapakoneta, Ohio, climbed slowly down a short ladder and reported in a steady voice that his footprint had depressed the soil only a fraction of an inch, that 'the surface appears to be very fine-grained as you get close to it, it's almost like a powder down there, it's very fine'.

Shortly before NASA launched Apollo 11, it received a letter from the Union of Persian Storytellers begging NASA to change the plan: a moon landing would rob the world of its illusions and rob the union's members of their livelihood. During the spacecraft's flight, the Mission Control Center in Houston asked the crew to look out for Chang'e and for her bunny, too. Houston said that the bunny would be 'easy to spot, since he is always standing on his hind feet in the shade of a cinnamon tree'. Buzz Aldrin responded, 'We'll keep a close eye out for the bunny girl.'

*

'The moon is hot again,' Jack Burns, the director of the NASA-funded Network for Exploration and Space Science (NESS), told me. NESS's headquarters are at the University of Colorado, Boulder, which has educated nineteen astronauts. (Boulder was also the setting for the television sitcom *Mork & Mindy*, in which Robin Williams played an alien from the planet Ork.) Part of NESS's mission is to dream up experiments to be done on the moon. An informational poster at the entrance reads 'Challenges of Measuring Cosmic Dawn with the 21-cm Sky-Averaged Global Signal'. In the decades since Apollo 11, NASA has invented Earth-mapping satellites, launched the Hubble Space Telescope, collaborated on the International Space Station and studied Mars. But none of these projects has generated the broad and childlike wonder of the moon.

Burns, who is in his mid-sixties, remembers the Mercury, Gemini and Apollo missions – the Cold War-era efforts, beginning in the late 1950s, that put men in space and finally landed them on the moon. He teaches a course on the history of space policy. 'The US had already lost the start of the space race,' he said of the origins of Apollo. 'The Soviet Union was first with a satellite in space. They were first with an astronaut in space.' Yuri Gagarin's journey into outer space took place in April 1961. President John F. Kennedy delivered his moon-shot speech the following month, and Congress eventually allocated 4.4 per cent of the national budget to NASA. 'But, if you

'Advances in engineering could turn the moon into a way station for launching rockets and satellites farther into the solar system, to Mars and beyond. (The weak gravity on the moon dramatically eases launches.)'

live by political motivations, you die by political motivations,' Burns said. 'Apollo died. Nixon killed the programme.' Only twelve people have walked on the moon, all of them between the summer of 1969 and Christmas 1972. All the moonwalkers were men, all were American, all but one were Boy Scouts and almost all listened to country-and-western music on their way to the moon; they earned eight dollars a day, minus a fee for a bed on the spacecraft. Since the last moonwalk, humans have launched crafts that have orbited the moon, crashed probes into it and taken increasingly detailed photos of it. But no one has been back.

The planetary scientist Bruce Hapke, who has a yellowish, opaque lunar mineral – hapkeite – named for him, said, 'Almost every president since Nixon proposed going back to the moon.' (President Obama focused instead on studying an asteroid near Earth and working towards the distant goal of sending astronauts to Mars.) 'But the money was never allotted. Congress decided we couldn't have guns and the moon at the same time.' The Department of Defense's budget is now nearly $700 billion, whereas NASA's funding is $21.5 billion, or around half of 1 per cent of the national budget. The USA is still believed to spend more on space programmes than the rest of the world combined. (China's budget, however, is unknown.) Hapke said, 'The trouble is, there was always some kind of

emergency, always some war going on. Though that Cold War mentality also got us to the moon.'

Hapke recalls being told by several scientists and NASA employees that, 'when the moon landing was first conceived, it was a strictly political stunt: go to the moon, plant the flag, and come back to Earth'. The original design of the spacecraft allotted little to no room for scientific payloads. 'When the scientific community got wind of this, they pointed out strongly to NASA all the fantastic science that could be done, and the whole tone of the project was changed,' he said. Hapke was then at Cornell, where he and his lab mates studied what the lunar soil might be like; the moon's characteristic reflectivity helped them deduce that the surface must be a fine dust. For Hapke, the Apollo era remains the most exciting time in his scientific life. He also recalls 'the widespread puzzlement in both Congress and the general populace after the first landing: "We beat the Russians. Why are we going back?"'

Burns said, 'This time we need a more sustainable set of goals and reasons' for going to the moon. He meant a science mission or a business mission or both. 'We don't like to say we're going back to the moon,' but forward, he added. 'Our objectives are different. Our technology is different. Apollo had five kilobytes of RAM. Your iPhone is millions of times

For future missions to the moon, you might be surprised to learn that dust is one of the experts' concerns. The moon's surface is covered with regolith, which contains sediments, fragments of material and moon dust. Because the moon has no atmosphere or magnetic field, the upper layer of its regolith is constantly bombarded with cosmic rays and the solar wind, giving the dust an electrostatic charge. US astronauts who visited the moon were obsessed with it – and for a good reason: the dust is extremely abrasive, with an incredible ability to stick to things, so it scratched the visors of their helmets, weakened the seals of their pressurised suits, irritated their eyes and gave some of them coughs and sinusitis. Anything smaller than ten microns tends to remain trapped in our lungs, and moon dust particles are less than two microns, as fine as flour. This is because, on impact with the ground, the frequent micrometeoroid showers create tiny pieces of glass that are never smoothed off by wind or water, making the dust on the ground extremely fine. Lunar regolith could also turn out to be a resource, however. A group of researchers at ESA has used simulated regolith samples to test a method of extracting oxygen using molten salt electrolysis. Regolith is 40–45 per cent oxygen, so a regolith oxygen plant would use a material available on the moon to make breathable air for inhabitants and, above all, to produce fuel (liquid oxygen is one of the main propellants used in rockets), perhaps to power craft that could take us further in our exploration of space.

more powerful.' Watching the footage of Neil Armstrong's first steps, it takes a moment for one's eyes to make sense of the low-resolution image, which could easily be overexposed film or a painting by American abstract expressionist Robert Motherwell. 'It's amazing they made it.'

Burns told me that advances in engineering could turn the moon into a way station for launching rockets and satellites farther into the solar system, to Mars and beyond. (The weak gravity on the moon dramatically eases launches.) Lunar construction projects now look feasible. 'Down the hall we have a telerobotics lab,' Burns said. 'You could print components of habitats, of telescopes. You use the lunar regolith' – the dust of the moon – 'as your printing material. You could print the wrench you need to fix something.' Two decades ago the moon was believed to be a dry rock; now we know that there's water there. Both private industry and national agencies regard the mining of water and precious materials as something that's not too far off. There's space tourism, too, although the quiet consensus among scientists seems to be that the idea is goofy and impractical (see 'Mass Tourism in Space' on page 175).

NASA would like to establish a permanent presence on the moon, using reusable rockets and landers. The agency is working on the largest, strongest, fastest – of course – rocket yet, but it plans to

Montage of photographs of the moon's surface taken during the Apollo 14 and Apollo 15 missions.
Original images: NASA

THE PASSENGER   Rivka Galchen

'Jack Burns likens this de facto government support of commercial space exploration to the dawn of the airline industry. "In the 1920s early airline companies survived only because the government paid them to deliver the mail … I think we're looking at something similar with space exploration."'

purchase other equipment, including rockets and landers, off the shelf, from commercial companies. Bob Jacobs, a spokesperson for NASA, told me, '85 per cent of NASA's budget is for commercial contracts. We build what only we can build; the other services we look to purchase from approved vendors.'

Burns likens this de facto government support of commercial space exploration to the dawn of the airline industry. 'In the 1920s early airline companies survived only because the government paid them to deliver the mail.' It wasn't until later, when ordinary people became aeronauts, that the airline industry became economically viable. 'I think we're looking at something similar with space exploration,' Burns said.

There are also more emotionally leveraged business models, like that of Celestis, a funeral-services company, which puts cremains into space and has plans to take them to the moon. The Japanese beverage Pocari Sweat wants to be the first sports drink on the moon. Its manufacturer has booked a spot on a lunar lander developed by a Pittsburgh-based company, Astrobotic, which is scheduled to launch in the fourth quarter of 2022 and to land in the Lacus Mortis – the Lake of Death, which is actually a dry, flat area. Pocari Sweat employees have collected stories of children's dreams from across Asia and etched them on to titanium plates. The plates will be put inside a capsule designed to look like a Pocari Sweat can and will travel with some Pocari Sweat powder that will one day – so the plan goes – be mixed with moon water.

Even in fantasy, space ventures have always mingled idealistic and worldly motives. H.G. Wells published *The First Men in the Moon* in 1901. The novel's narrator, Mr Bedford, wants to make money. His collaborator, Mr Cavor, dreams of knowledge. Together they go to the moon. When they encounter moon dwellers – 'compact, bristling' creatures, 'having much of the quality of a complicated insect' – Bedford wants to destroy them; Cavor wants to learn from them. Bedford finds gold and embarks 'upon an argument to show the infinite benefits our arrival would confer upon the moon', involving himself 'in a rather difficult proof that the arrival of Columbus was, on the whole, beneficial to America'. Cavor is indifferent to the gold – it's a familiar mineral. Moon dwellers capture and chain Bedford and Cavor, then march them underground. Cavor assumes that there must be other, less brutal moon dwellers, as enlightened and knowledge-loving as he. In the end, Bedford makes it back to Earth. Cavor is presumed dead. But no one with a heart reads the novel and wants to be Bedford.

Burns grew up in Shirley, Massachusetts. Neither of his parents graduated from high school. From the age of five he knew that he wanted to study the stars. When I asked him what he hopes to see on the moon, he became suddenly boyish. 'I'd love to set up a low-frequency radio telescope on the far side of the moon, free from the interference of Earth signals. It could see to the beginnings of time. And the far side of the moon has craters there that were formed during the Late Heavy Bombardment four billion years ago.' During the Late Heavy Bombardment, large numbers of meteors crashed into the inner solar system. The period coincides roughly – and perhaps not coincidentally – with the beginnings of life on Earth. Burns said, 'Earth was also bombarded, but here that history has been erased or buried by weather, erosion. On the moon it's still right there on the surface. It's a history book. I'd like to read that book.'

*

The night I met with Burns was the eve of a supermoon – when the moon is both full and as close to Earth as it gets. I walked over to the Sommers-Bausch Observatory, not far from Burns's office; there was a bunny in the bushes, trying not to be noticed. Carla Johns, who operates the observatory's telescopes, met me in the hallway, which is lit in red to keep your eyes adapted to the dark. On the top floor she pressed a button, and the roof noisily rolled back. There it was, with all its starry friends. Johns explained how the telescopes worked – they are essentially buckets of light. She said that children often shout when they see the moon so close.

Johns showed me a collection of small telescopes, and discussed the 18th-century French astronomer Charles Messier. 'Back then, the way astronomers made money was finding comets and telling kings they had a comet to name after them,' she said. When Messier was eleven, his father died, and afterwards he received no formal schooling. But he developed an exceptional gift for finding comets. 'To find those comets, he documented everything he could see in the sky,' Johns said. 'Once he was sure a sky object wasn't a comet, it was of no interest to him. Some of that stuff he found turned out to be Andromeda and the Crab Nebula.' She showed me a large telescope on a mount developed by John Dobson, a chemist by training, who worked briefly on the Manhattan Project, then resolved to spend the rest of his life as a monk. While living at a monastery in San Francisco, he would walk the shipyards, gathering old porthole glass to fashion into homemade telescopes, which he would share with others in sidewalk astronomy lectures. 'The monks eventually asked him to leave,' Johns said.

Johns became a telescope operator relatively late in her professional life. She had worked in human resources, and enjoyed it, but at a difficult moment she found herself at the Denver Museum of Nature & Science, where her parents used to take her as a child. 'I looked through the telescope, and I began to cry,' she said. She had always loved science but had chosen another career because of family and financial issues. 'I said to myself, "I need to be involved with this."'

*

Shortly before the turn-off for the town of Mojave, California, there were train cars along the right side of the road, painted old-fashioned black and standing

Processed image of a section of a 'relatively recent' moon crater photographed by Apollo 15.
Original images: NASA

THE PASSENGER   Rivka Galchen

still. On the left were hundreds of white wind turbines, spinning. Soon I came to a slightly weathered sign for the Mojave Air and Space Port – 'Imagination Flies Here' – which features a picture of a young boy holding a toy plane. You're allowed to launch rockets here; you're allowed to fly objects beyond the atmosphere. A number of aerospace firms have offices at the port.

In November 2018 NASA named nine companies to be part of its Commercial Lunar Payload Services programme: if NASA wants to send something to the moon, these companies are approved to provide transportation. (Five more companies were added to the programme in November 2019.) 'FedEx to space', I was told to think of it. 'Or DHL'. Some of them are large and well known, like Lockheed Martin Space. Masten Space Systems has sixteen employees. It is based at the Air and Space Port, down the road from Virgin Galactic, in offices that resemble the extra building my elementary school put in the playground when enrolment exceeded capacity. When Masten won a NASA-funded prize – for vertical take-off and precision landing in conditions simulating those of the moon – it had five employees. Its winning rocket, Xoie, looks like a slim, silvery water tower, only thirty centimetres tall – two stacked spheres on a tripod, with tanks of helium on the sides.

'Our focus is on reusable rockets,' Masten's CEO Sean Mahoney told me. 'We have a rocket that has flown two hundred and twenty-seven times. We want space to be affordable.' Masten plans to begin taking payloads to the moon in 2023, 'Mostly science payloads, mostly NASA. Some commercial.' Among the items that NASA wants to send are a solar-power cell and a navigation device that the agency will test in lunar conditions.

Mahoney and I talked over a meal at the Voyager Restaurant, on the grounds of the spaceport. The Voyager looks like Mel's Diner from the TV show *Alice*. (A lot about lunar exploration reminded me of old television shows, especially *Bonanza*.) I had a grilled-cheese sandwich – spaceport food. Mahoney said, 'There's the PBS version of space, which is beautiful. And that is real. But, also, space – well, you've heard of the military-industrial complex? Space is an offshoot of that.' Something shiny and fleet was taking off in the distance, and the windows shook. Mahoney pointed out a tumbleweed blowing across the lot. 'I'm a business guy by background, not a space guy, so I had to learn all of this,' he said. Mahoney believes that, because the space industry was a government-sanctioned monopoly for decades, there was no room for risk or for competition; the fear of failure dominated. 'Lockheed Martin and Boeing could charge exorbitant prices,' he said. 'As a business person, when you see a fat margin – when you see a service that can be provided much more cheaply – you see value.'

We walked through strong winds to the hangars where Masten does its manufacturing. There were none of the vacuum chambers and clean white rooms that one associates with rocket science. Instead, there were trailer beds loaded with rocket parts for testing; there were purple-and-yellow long-sleeved T-shirts for launch days. There were tanks of helium, wrenches of every size. A young man wearing an Embry-Riddle Aeronautical University sweatshirt and a welding mask was making an engine casing.

Mahoney pointed out an engine without its casing, next to a small computer. 'Some of these rocket models are literally operated by Raspberry Pi,' he said.

'Raspberry *pie*?'

'That's a very basic computer. A $35 computer. My point being, some of our parts we can buy at Home Depot.'

Masten was founded in 2004 by David Masten, a former software-and-systems engineer, who remains the chief technical officer. 'When I was a kid, I was going to be an astronaut,' Masten told me. 'But, by the 1980s, space was getting boring – it wasn't going anywhere – and there was this new thing called computers.' He became an IT consultant and eventually worked at a series of startups. Throughout, Masten's hobby remained rockets. 'My thought was that, maybe, instead of doing the heavy analysis traditional of the aerospace industry, you do something more like I was used to,' he said. 'You write some code, you compile it, you test it and you iterate over and over in a tight, rapid fashion. I wanted to apply that method to rocketry.'

When a Masten rocket takes off, it has a delicate appearance. One of the newer ones, the Xodiac, looks like two golden balloons mounted on a metal skeleton. A kite tail of fire shoots out as the Xodiac launches straight up; at its apex, it has the ability to tilt and float down at an angle, as casually as a leaf. When Xodiac nears its designated landing spot, it abruptly slows, aligns, seems to hesitate, lands. It's eerie – at that moment, the rocket seems sentient, intentional.

In one demonstration, the Xodiac performed a deceptively mundane task: it carried a 'planetvac' – an invention intended to vacuum dust from the lunar surface – up and over one metre, deployed the vacuum, then scooted up and over another metre, hopping like a lunar janitor. The rockets are self-guided, unless overridden by a human; they are doing their own thing. 'We believe computers can fly rockets better than people can,' Masten told me.

Water will be indispensable to any future moon bases, above all as a raw material for the fuel to power any rockets launched from there. In November 2008 India used a probe from the satellite *Chandrayaan-1* to document the presence of water in the lunar atmosphere. The following year the same satellite recorded the presence of hydroxyl fragments. Two studies published in *Nature Astronomy* in October 2020 then showed that water might be more accessible on the moon than previously believed. The surface area that remains in perpetual darkness, where the presence of ice is more likely, is larger than had been thought. There are claimed to be around forty thousand square kilometres of 'cold traps' on the moon, most of them near the poles, but there are also some at lower latitudes, inside craters or depressions where the shadows keep temperatures very low. The other study, coordinated by NASA, used the SOFIA infrared telescope mounted on a Boeing 747SP to confirm the presence of water molecules in quantities ranging from one to four parts per ten thousand. How useful that water might be depends above all on how it is stored. It could be in the form of ice crystals – in which case lunar settlers could simply heat the regolith to extract its water – but it is more likely that the water is trapped in minuscule glass beads that form when the lunar surface is hit by micrometeorites, and extracting water from these beads would not be an easy task.

'Fuel accounts for around 90 per cent of the weight of a rocket, and every kilogram of weight brought from Earth to the moon costs roughly $35,000; if you don't have to bring fuel from Earth, it becomes much cheaper to send a probe to Jupiter.'

Many scientists see little need for humans on the moon, since robots would do the work more safely and inexpensively.

*

'Now, you will ask me what in the world we went up on the moon for,' Qfwfq, the narrator of Italo Calvino's *Cosmicomics*, says. 'We went to collect the milk, with a big spoon and a bucket.' In our world, we are going for water. 'Water is the oil of space,' George Sowers, a professor of space resources at the Colorado School of Mines, in Golden, told me. On the windowsill of Sowers's office is a bumper sticker that reads 'My other vehicle explored Pluto'. This is because his other vehicle *did* explore Pluto. Sowers served as the chief systems engineer of the rocket that, in 2006, launched NASA's *New Horizons* spacecraft, which has flown by Pluto and carried on to Ultima Thule, a snowman-shaped, thirty-kilometre-long rock that is the most distant object a spacecraft has ever reached. 'I only got into space resources in the past two years,' he said. His laboratory at the School of Mines designs, among other things, small vehicles that could one day be controlled by artificial intelligence and used to mine lunar water.

Water in space is valuable for drinking, of course, and as a source of oxygen. Sowers told me that it can also be transformed into rocket fuel. 'The moon could be a gas station,' he said. That sounded terrible to me, but not to most of the scientists I spoke to. 'It could be used to refuel rockets on the way to Mars' – a trip that would take about nine months – 'or considerably beyond, at a fraction of the cost of launching them from Earth,' Sowers said. He explained that launching fuel from the moon rather than from Earth is like climbing the Empire State Building rather than Mt Everest. Fuel accounts for around 90 per cent of the weight of a rocket, and every kilogram of weight brought from Earth to the moon costs roughly $35,000; if you don't have to bring fuel from Earth, it becomes much cheaper to send a probe to Jupiter.

Down the hall, in the Center for Space Resources' laboratory, near buckets of lunar and asteroid simulants, was a small 3D printer. Four graduate students were assembled there with Angel Abbud-Madrid, the centre's director. I asked them how difficult it would be to 3D-print, say, an electrolyser – the machine needed to separate the hydrogen and oxygen in water to make rocket fuel. They laughed.

'Here, let me show you something very fancy,' Hunter Williams, who was wearing sapphire-coloured earrings, said. He poured some Morton sea salt into a plastic cup and added water. He stuck two silver thumbtacks through the bottom of the plastic cup, then held a battery up to them. Small bubbles began forming on the thumbtacks. The oxygen was separating from the hydrogen. You probably did this

A Brief Guide to Achieving Orbit on a Budget

Reusable rockets

The main reason we have yet to colonise the solar system is the prohibitive cost of leaving Earth. To send just half a kilo of stuff into low Earth orbit costs on average $10,000. This is because a rocket – the only way we have managed to reach space so far – is essentially a large tube of propellant with a small cargo on top, comprising around 4 per cent of its total mass. The propellant in itself is not that expensive, but the rocket, which is almost always discarded, really is. So recycling the rocket can produce major savings, and that is what SpaceX and other companies are attempting to do. The Falcon 9 and Falcon Heavy are already partly reusable, and sending things into orbit with SpaceX costs a lot less than with other operators as a result (see the infographic on page 6) – and videos of it landing vertically are pretty cool. Elon Musk is also planning a completely reusable spacecraft known as the Starship. But even if you reuse parts of the rocket or the boosters, rockets remain highly inefficient: the bulk of the energy produced by the propellant is used to transport the propellant itself from Earth into orbit. If we could eliminate this problem we would save a whole lot of cash.

Starting from on high

One solution would be to start from a high altitude, which would save you carrying the propellant required to get up there. But the problem is not so much the altitude as the speed. In spite of what one might think when watching astronauts floating around in the ISS, gravity at that altitude is about 90 percent of what we experience on Earth. Astronauts float not because there is no gravity but because they are in free fall – at the same time, they are travelling extremely fast, at a speed of 7.7 km/second. Imagine firing a cannonball from a tower: the faster the cannonball, the further away it will fall. Above a certain speed, rather than falling, it will go into orbit. So the difficulty lies in getting it to the required speed. The idea, used by Virgin Orbit, is to get as high and as fast as possible using an aircraft (a modified Boeing 747 named *Cosmic Girl*) and release the rocket from there – with the advantage that you can take off from any airport and there is no need to worry about the weather. But the altitude and speed gained thanks to the aeroplane are a small percentage of what you need, and the size of the rocket will necessarily be limited: the method is viable – *Cosmic Girl* has already successfully launched its first rocket – but the costs are not drastically lower.

The extremely long-range super-cannon

So why not use a cannon then? This is not a completely unrealistic idea and, in fact, was seriously considered in the 1950s and 1960s by the USA and Canada, which funded Project HARP (the High Altitude Research Project). The advantage is clear: there is no need to carry propellant, and the equipment can be reused endlessly and with a quick turnaround. The disadvantages are equally clear: first, the cost of constructing a cannon with a barrel three metres across and over a hundred or so metres long. To be usable multiple times, it would also need a chamber able to cope with repeated powerful explosions. Another issue is that the acceleration imparted by the explosion would instantly kill a human being, so it would only suit non-live payloads. Project HARP successfully fired a projectile to the remarkable altitude of 179 kilometres, but by that point NASA had opted for rockets, and the programme was closed down. The story does not end there, however. Canadian scientist Gerald Bull, one of the architects of Project HARP, ended up in Iraq on the payroll of Saddam Hussein working on the Project Babylon super-cannon. He was later killed outside his apartment in Brussels, some say by Mossad agents.

Space elevators

Talking of methods that could double up as extremely powerful weapons, we should mention the laser ignition system for liquid oxygen and hydrogen. But we won't, for two reasons: we have yet to invent a laser powerful enough and, even if we did, it is more likely that we would use it to incinerate one another. So we will move on to the most elegant and seductive method: the space elevator, dreamed up by Konstantin Tsiolkovsky (see 'The Cosmic Comedy' on page 77). The idea is a simple one: from a point on the Earth's equator you run a cable up to a space station in geostationary orbit (at an altitude of thirty-six thousand kilometres, where it would rotate around the Earth in precisely one day, apparently maintaining a fixed position in the sky). You also need a counterweight beyond the space station, even further out into space, to keep the cable under tension. And then you run a vehicle up and down the cable to transport your cargo and – why not? – people as well. Simple, right? Unfortunately, we have yet to invent a material strong and light enough for the cable, which has to be able to support its own weight. There are also other issues: weather disturbances, oscillation of the cable, satellites and space junk in orbit to avoid. It seems unlikely that this marvellous project will be built on Earth. On the other hand, it would stand more of a chance on the moon, where the lower gravity and absence of an atmosphere would make things easier. The operators would, of course, also have to work on improving the elevator Muzak.

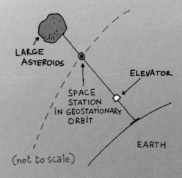

LARGE ASTEROIDS

ELEVATOR

SPACE STATION IN GEOSTATIONARY ORBIT

EARTH

(not to scale)

experiment in middle school without knowing that you were doing rocket science. 'The idea is for whatever goes up to the moon to be that simple,' Williams said. 'To be that basic.'

'It would be like living off the land,' Ben Thrift, another graduate student, added. Thrift studied theatre as an undergraduate, and later ran a bakery before earning a degree in engineering and enrolling in the space-resources programme. 'I decided to grow up and do something real,' he said.

'By "real" he means go to the moon,' Abbud-Madrid said.

*

'Transportation is not an end in itself,' Sowers told me. He is excited about solar power, which already runs many satellites in space, where there is no night or clouds. He speculates that if we had a base on the moon we could use 3D printers to make giant solar panels, as large as two kilometres across, which could be launched into orbit; the resulting power could be beamed back to Earth via microwave radiation. 'Space solar would be an unlimited, inexhaustible source of green energy,' Sowers said. 'It requires no magic, and much of the technology is ready. I think we could do it by 2030, if we wanted to.' Another bumper sticker in Sowers's office reads 'Physicists have strange quarks'.

Other specialists have a different view of the resources available in space. Asteroids contain precious metals, such as platinum, palladium and gold. A number of asteroid-mining companies have come and gone since 2015, when Neil deGrasse Tyson remarked that 'the first trillionaire there'll ever be is the person who mines asteroids for their natural resources'. But asteroid hunting is like whaling, in the length of its missions and the speculative

DEEP-SPACE MINES

Until recently asteroids were seen as existential threats, and we have studied their trajectories in the hope of never coming across them. But now we are actively seeking them out ... We have discovered that these curious objects, found in huge numbers in our solar system (there are over a million of them), are extremely rich in such precious metals as platinum, iron, nickel, titanium and cobalt. According to some estimations, just one asteroid, 16 Psyche (discovered in 1852 at the Astronomical Observatory of Capodimonte in Naples), would be worth more than the entire global economy. With a diameter of 250 kilometres, it is composed entirely of pure iron and nickel. But, to start with, we could make do with one of its smaller neighbours, 3554 Amun, measuring just two kilometres across; this would provide us with thirty times the amount of all the metals ever extracted on Earth. As a result, companies financed by the usual suspects, including Branson and Bezos (see 'The Space Barons' on page 142), have been studying the feasibility of mining in space for years, looking for ways to safely land the heaven-sent bounty here on Earth. Some are looking at how to use all these resources in situ, transforming meteorites and other astronomical objects into outposts for the conquest of the universe. In this scenario, robotics and 3D printers could be used to turn the metals into equipment and habitats, significantly reducing costs compared with launching them from Earth. Others contend that the principal resource when it comes to space exploration is water, which will provide oxygen (to breathe) and hydrogen (for rocket fuel). But whether they are intergalactic service stations or resources to be pillaged, one thing is for sure: asteroids will continue to be a hot topic for the foreseeable future.

nature of its success; the moon is only three days away, and its movements are extremely well known to us. NASA recently named ten companies as potential contractors for equipment to gather and analyse soil in space.

One of them was Honeybee Robotics. I visited its exploration-technology division in Pasadena, which, from the outside, looks as dull as ditchwater, a collection of beige concrete buildings. Inside were lunar-rock samplers, the planetvac that was tested on a Masten rocket, some Nerf guns – and WINE (which stands for 'the World Is Not Enough'), a steam-powered spacecraft designed to find water in lunar dirt (or on asteroids), convert it to energy then hop to the next site to pull up samples and more water for fuel.

Kris Zacny, a vice-president of Honeybee Robotics, was expecting his third child in the next few days. 'So much has to do with where you're born,' he said, explaining how he came to the field of space mining. Zacny is originally from Poland, the son of a musician father, who wanted him to be a musician as well. 'What a disappointment I must have been,' Zacny said. 'I spent my time thinking about the moon.' When he was seventeen, his family moved to South Africa. Zacny went to college on a scholarship from De Beers and worked in the diamond mines while in school. 'I graduated top of my class, with a degree in mechanical engineering, and next thing I knew I was

twelve thousand feet underground,' he said. He spent two years in a coal mine, and a month in a gold mine that at the time was the deepest mine in the world. 'I always dreamed of space, but it wasn't an option for me,' he said.

In 2000 he landed a one-year position as a research assistant for a professor in Berkeley's Materials Science and Engineering Department. 'I knew it was too late for me to be a space guy, I accepted that. But I had the mining expertise. I said to the professor, "Don't laugh at me, but I'd like to do extraterrestrial mining."' What can be found on the moon remains for the most part unknown, although there is reasoned speculation. Honeybee is one of a growing number of companies that are developing standardised lunar rovers. Small countries with no national space agency, as well as private entities, could soon have their own robotic resource hunters roving around on the moon with little honeycomb emblems on their sides.

*

Buzz Aldrin had hoped, and briefly expected, that it would be he, and not Neil Armstrong, who would take the first human step on the moon. The astronaut Michael Collins, who manned the control module that orbited the moon while Armstrong and Aldrin walked below, has said of Aldrin that he 'resents not being first on the moon more than he

'A number of asteroid-mining companies have come and gone since 2015, when Neil deGrasse Tyson remarked that "the first trillionaire there'll ever be is the person who mines asteroids for their natural resources".'

'The guiding laws of space are defined by the Outer Space Treaty, from 1967, which has been signed by 108 countries, including all those with substantial space programmes.'

appreciates being second'. On the moon, Armstrong took photos of Aldrin posing, but Aldrin took none of Armstrong doing the same. One of the few photos that shows Neil Armstrong on the moon was taken by Armstrong himself – of his reflection in Aldrin's helmet as Aldrin salutes the flag. We are petty and misbehave on Earth; we will be petty and misbehave in space.

The guiding laws of space are defined by the Outer Space Treaty, from 1967, which has been signed by 108 countries, including all those with substantial space programmes. 'Laws that govern outer space are similar to the laws for the high seas,' Alain Berinstain, the vice-president of global development at the lunar-exploration company Moon Express, explained. 'If you are two hundred miles away from the continental shelf, those waters don't belong to anybody – they belong to everybody.' Moon Express describes the moon as the eighth continent. The company, which is based in Florida, had to postpone the delivery of its first lander to the moon, with telescopes and the Celestis cremains on board, following legal and financial problems. 'If you look down at the waters from your ship and see fish, those fish belong to everybody,' Berinstain continued. 'But, if you put a net down and pull those fish on to the deck of the ship, they're yours. This could change, but right now that is how the US is interpreting the Outer Space Treaty.'

Individual countries have their own interpretations of the treaty and set up their own regulatory frameworks. Luxembourg promotes itself as 'a unique legal, regulatory and business environment' for companies devoted to space resources and is the first European country to pass legislation similar to that of the US, deeming resources collected in space to be ownable by private entities (see 'The Grand Duchy of Space' on page 52).

It's not difficult to imagine moon development, like all development, proceeding less than peacefully and less than equitably. (At least, unlike with colonisation on Earth, there are no natives whose land we're taking, or so we assume.) Philip Metzger, a planetary physicist at the University of Central Florida, said, 'I'm really glad that all these countries, all these companies, are going to the moon. But there will be problems.' Any country can withdraw from the Outer Space Treaty by giving a year's notice. 'If any country feels it has a sufficient lead in space, that is a motivation to withdraw from the treaty,' he said.

So there is a tacit space race already. On the one hand, every national space agency applauded the success of the *Chang'e-4* lander. The mission had science partnerships with Germany, the Netherlands, Saudi Arabia and Sweden. NASA collaborates with many countries in space, sharing data, communications networks and expertise. Russian rockets

A photomontage of a view of the dark side of the moon close to Crater No. 308, taken during the Apollo 11 mission, and microscope images of grains of lunar soil collected by Apollo 17. Original images: NASA

bring American astronauts to the International Space Station (ISS). When, in response to economic sanctions in 2014 following Russia's annexation of Crimea, the head of the Russian space agency said that maybe the American astronauts could get to the ISS by trampoline, the comment was dismissed as posturing. Still, NASA contracted with SpaceX, Elon Musk's rocket company, to take astronauts to the ISS, and the first flight took off on 30 May 2020, making the USA less reliant on Russia. There were further tensions after February 2022 in the wake of economic sanctions following Russia's invasion of Ukraine, and fears that NASA astronaut Mark Vande Hei might be left behind on the ISS rather than be brought home by a Russian Soyuz spacecraft, but NASA stressed that joint Russian–US operations on the ISS were continuing.

Meanwhile Russia and China announced on 9 March 2021 that they will work together on a moon base. NASA used to collaborate with the China National Space Administration (CNSA), but eleven years ago, in 2011, just six months after members of NASA visited the CNSA, Congress passed a bill that effectively prohibited collaboration.

It's natural to want to leave the moon undisturbed; it's also clear that humanity will disturb it. But do we need to live there? Jeff Bezos, the founder of Amazon, envisages zoning the moon for heavy industry and Earth for light industry and residential purposes. Bezos's company Blue Origin is developing reusable rockets intended to bring humans reliably back and forth from space, with the long-term goal of creating manufacturing plants there, in zero gravity. Earth would be eased of its industrial burden, and the lower-gravity conditions would be beneficial for making certain goods, such as fibre-optic cables.

THE GRAND DUCHY OF SPACE

Luxembourg is a country about the size of a large asteroid. It has no raw materials and not much tourism, but it does have one resource that it is determined to exploit: national sovereignty. While old and new powers, as well as smaller petrostates such as the UAE, continue to invest money in space agencies, Luxembourg has opted for a different path, which is to become the base on Earth for companies that want to go into space to make money. As it has done in the past by adopting a tax system favourable to multinationals and investment funds, the Grand Duchy is transforming itself into a 'space haven' by creating a legal and commercial structure suited to the exploitation of space and supporting companies active in the NewSpace sector. In 2017 its parliament passed a law guaranteeing that any company based in the country would have legitimate ownership of any resources extracted from any astronomical object. It was the first law of its kind in Europe and, unlike the American Space Act passed by the Obama administration in 2015, it is not limited to companies owned by Luxembourgers. In 2020 the government invested in a fund, Orbital Ventures, as a way of financing innovative NewSpace startups and approved a five-year national action plan for space science and technology. These initiatives are partly a reaction to the tax revelations in the 2014 LuxLeaks scandal, which forced the Grand Duchy to clean up its image and to find a new way to use its sole precious resource.

'There's the argument that we've destroyed the Earth and now we're going to destroy the moon. But I don't see it that way,' Metzger said. 'The resources in space are billions of times greater than on Earth. Space pretty much erases everything we do. If you crush an asteroid to dust, the solar wind will blow it away. We can't really mess up the solar system.'

*

The most likely origin story for the moon is that it was formed 4.5 billion years ago after a Mars-size planet called Theia crashed into Earth. Theia broke into countless pieces, which orbited Earth. Slowly – or quickly, depending on your timescale – the shards coalesced and formed the moon we know today, the one that is drifting away from us at a rate of four centimetres or so per year. If we had two moons, like Mars does, or sixty-two, like Saturn, we wouldn't feel the same way about our moon.

Zou Xiaoduan, a scientist who worked on all phases of the Chang'e project, was born in 1983 in Guizhou province in south-west China – 'a very poor place back then', she told me. As a child, she said, she 'was stunned to learn that the moon was not a weird monster following me around'. She remembers hearing her family chatting about the Apollo missions. That men had been on the moon seemed unfathomable to her. She asked every adult to confirm it. She wanted to become an astronaut – a goal she attributes to there not being any Disney movies for her to watch. She began work on China's lunar programme in 2006. 'I still recall the first lunar image from *Chang'e-1* being shown to me,' she said of the images sent home in 2007, during China's first lunar orbital mission. 'And the first time *Chang'e-2* flew by an asteroid, 4179 Toutatis,' three years later,

'no one had ever seen that asteroid.' Zou came to the USA in 2015 and now works for the Planetary Science Institute in Tucson. She is part of a mission studying the asteroid Bennu, which NASA describes as 'an ancient relic of the solar system's early days'. Like everyone else I spoke to who studies the moon, she loves her job. Of her work on the Chang'e missions, she said that every image has been 'thrilling, every moment is a "wow"'. She continued, 'I'm just so excited and super happy that I picked this career.'

The twelve men who walked on the moon, who saw Earth as a distant object – did they lose their illusions? A couple had alcohol problems, one co-founded the Institute of Noetic Sciences and one became an evangelical preacher. One became a one-term Republican senator who has denied that humans are responsible for climate change; another became a painter, of the moon. Neil Armstrong was one of the few who had a mostly steady, unremarkable post-moonwalk life. He moved to a dairy farm and became a professor at the University of Cincinnati. Nearly a decade after his trip to the moon he wrote a poem called 'My Vacation':

Nine Summers ago, I went for a visit.
To see if the moon was green cheese.
When we arrived, people on earth
    asked: 'Is it?'
We answered: 'No cheese, no bees, no
    trees.'

There were rocks and hills and a
    remarkable view
Of the beautiful earth that you know.
It's a nice place to visit, and I'm certain
    that you
Will enjoy it when you get to go. 🐦

What Happens if China Makes First Contact?

ROSS ANDERSEN

The FAST (Five-hundred-metre Aperture Spherical Telescope) radio telescope in the karst mountains of south-western China, is the largest and most sensitive in the world. Construction was completed in 2016, and it became fully operational in 2020.
Credit: Visual China Group/Getty Image

While the United States seems to have turned its back on the search for extraterrestrial intelligence, China has built the world's largest radio telescope with the aim of picking up signals from distant worlds. But if we look at how past encounters between different civilisations on Earth have played out, is it really wise to announce our presence to the cosmos?

55

As the United States has turned away from searching for extraterrestrial intelligence, China has built the world's largest radio dish for precisely that purpose.

In 2017 the Chinese Academy of Sciences invited Liu Cixin, China's preeminent science-fiction writer, to visit its new state-of-the-art radio dish in the country's south-west. Almost twice as wide as the dish at the US National Science Foundation's Arecibo Observatory in the Puerto Rican jungle, the new Chinese dish is the largest in the world, if not the universe. Although it is sensitive enough to detect spy satellites even when they're not broadcasting, its main uses will be scientific, including an unusual one: the dish is Earth's first flagship observatory custom-built to listen for a message from an extraterrestrial intelligence. If such a sign comes down from the heavens during the next decade, China may well hear it first.

In some ways, it's no surprise that Liu was invited to see the dish. He has an outsize voice on cosmic affairs in China, and the government's aerospace agency sometimes asks him to consult on science missions. Liu is the patriarch of the country's science-fiction scene. Other Chinese writers I met attached the honorific *Da*, meaning 'big', to his surname. In years past, the academy's engineers sent Liu illustrated updates on the dish's construction along with notes saying how he'd inspired their work.

But in other ways Liu is a strange choice to visit the dish. He has written a great deal about the risks of first contact. He has warned that the 'appearance of this Other' might be imminent, and that it might result in our extinction. 'Perhaps in ten thousand years, the starry sky that humankind gazes upon will remain empty and silent,' he writes in the postscript to one of his books, *The Three-Body Problem* (Tor Books, 2014, USA / Head of Zeus, 2015, UK). 'But perhaps tomorrow we'll wake up and find an alien spaceship the size of the Moon parked in orbit.'

In recent years Liu has joined the ranks of the global literati. *The Three-Body Problem* became the first work in English translation to win the Hugo Award, science fiction's most prestigious prize. Barack Obama told *The New York Times* that the book – the first in a trilogy – gave him cosmic perspective during the frenzy

ROSS ANDERSEN is deputy editor of *The Atlantic*. He previously worked for *Aeon* magazine and the *Los Angeles Review of Books*, where he edited the science, technology and health section. He has won several awards for his long-form articles covering science, philosophy, history and culture and has been published by *The Economist* and *Scientific American* among others. He is working on a book on the search for extraterrestrial intelligence that will be published by Penguin Random House.

'No civilisation should ever announce its presence to the cosmos, says Liu Cixin. Any other civilisation that learns of its existence will perceive it as a threat to expand, as all civilisations do.'

of his presidency. Liu told me that Obama's staff asked him for an advance copy of the third volume.

At the end of the second volume, one of the main characters lays out the trilogy's animating philosophy. No civilisation should ever announce its presence to the cosmos, he says. Any other civilisation that learns of its existence will perceive it as a threat to expand, as all civilisations do, eliminating their competitors until they encounter one with superior technology and are themselves eliminated. This grim cosmic outlook is called 'dark-forest theory', because it conceives of every civilisation in the universe as a hunter hiding in a moonless woodland, listening for the first rustlings of a rival.

Liu's trilogy begins in the late 1960s, during Mao's Cultural Revolution, when a young Chinese woman sends a message to a nearby star system. The civilisation that receives it embarks on a centuries-long mission to invade Earth, but she doesn't care; the Red Guard's grisly excesses have convinced her that humans no longer deserve to survive. En route to our planet, the extraterrestrial civilisation disrupts our particle accelerators to prevent us from making advances in the physics of warfare, such as the one that brought the atomic bomb into being less than a century after the invention of the repeating rifle.

*

Science fiction is sometimes described as a literature of the future, but historical allegory is one of its dominant modes. Isaac Asimov based his *Foundation* series on classical Rome, and Frank Herbert's *Dune* borrows plot points from the past of the Bedouin Arabs. Liu is reluctant to make connections between his books and the real world, but he did tell me that his work is influenced by the history of Earth's civilisations, 'especially the encounters between more technologically advanced civilisations and the original settlers of a place'. One such encounter occurred during the 19th century, when the Middle Kingdom of China, around which all of Asia had once revolved, looked out to sea and saw the ships of Europe's seafaring empires, whose ensuing invasion triggered a loss in status for China comparable to the fall of Rome.

When I travelled to China to visit its new observatory, I first met up with Liu in Beijing. By way of small talk, I asked him about the film adaptation of *The Three-Body Problem*. 'People here want it to be China's *Star Wars*,' he said, looking pained. The pricey shoot ended in mid-2015, but the film has been postponed indefinitely. At one point, the entire special-effects team was replaced. 'When it comes to making science-fiction movies, our system is not mature,' Liu said.

I had come to interview Liu in his capacity as China's foremost philosopher of first contact, but I also wanted to know what to expect when I visited the new dish. After a translator relayed my question, Liu stopped smoking and smiled.

'It looks like something out of science fiction,' he said.

*

A week later I rode a bullet train out of Shanghai, leaving behind its purple *Blade Runner* glow, its hip cafés and craft-beer bars. Rocketing along an elevated track, I watched high-rises blur by, each a tiny honeycomb piece of the rail-linked urban megastructure that has recently erupted out of China's landscape. China poured more concrete from 2011 to 2013 than the United States did during the entire 20th century. The country has already built rail lines in Africa, and it hopes to fire bullet trains into Europe and North America, the latter by way of a tunnel under the Bering Sea.

The skyscrapers and cranes dwindled as the train moved farther inland. Out in the emerald rice fields, among the low-hanging mists, it was easy to imagine ancient China – the China whose written language was adopted across much of Asia; the China that introduced metal coins, paper money and gunpowder into human life; the China that built the river-taming system that still irrigates the country's terraced hills. Those hills grew steeper as we went west, stair-stepping higher and higher until I had to lean up against the window to see their peaks. Every so often, a Hans Zimmer bass note would sound, and the glass pane would fill up with the smooth, spaceship-white side of another train, whooshing by in the opposite direction at around 300 km/h.

It was mid-afternoon when we glided into a sparkling, cavernous terminal in Guiyang, the capital of Guizhou, one of China's poorest, most remote provinces. A government-imposed social transformation appeared to be under way. Signs implored people not to spit indoors. Loudspeakers nagged passengers to 'keep an atmosphere of good manners'. When an older man cut in the cab line, a security guard dressed him down in front of a crowd of hundreds.

The next morning I went down to my hotel lobby to meet the driver I'd hired to take me to the observatory. Two hours into what was supposed to be a four-hour drive, he pulled over in the rain and waded thirty metres into a field where an older woman was harvesting rice to ask for directions to a radio observatory more than 150 kilometres away. After much frustrated gesturing by both parties, she pointed the way with her scythe.

We set off again, making our way through a string of small villages, *beep-beep*ing motorbike riders and pedestrians out of our way. Some of the buildings along the road were centuries old, with upturned eaves; others were freshly built, their residents having been relocated by the state to clear ground for the new observatory. A group of the displaced villagers had complained about their new housing, attracting bad press – a rarity for a government project in China. Western reporters took notice: 'China Telescope to Displace 9,000 Villagers in Hunt for Extraterrestrials' read a 2016 headline in *The New York Times*.

*

The search for extraterrestrial intelligence (SETI) is often derided as a kind of religious mysticism, even within the scientific community. A quarter of a century ago the United States Congress defunded America's SETI programme with a budget amendment proposed by Senator Richard

Liu Cixin, born in 1963 in Yangquan, was the first Asian writer to win the prestigious American Hugo Award for Best Novel, in 2015, for *The Three-Body Problem*. Credit: Zachary Bako/Redux/Contrasto

In the few years since the inauguration of the Five-hundred-metre Aperture Spherical Telescope (FAST), Pingtang County has undergone development on a scale no less impressive than its vast radio telescope. One of the reasons for building the so-called 'eye of heaven' in this poor region of China was to attract investment, scientists ... and tourists. And, boy, did it work! Fifteen kilometres away, the town of Kedu, where, until recently, the only thing to aspire to was to get out, has been transformed into 'Astronomy Town': a complex of convention centres, tourist attractions, museums and IMAX cinemas that have sprung up out of nowhere, linked to the provincial capital by brand-new motorways. When you walk around at night your footsteps leave a trail of light on the pavement like a comet, above your head there are streetlights shaped like Saturn with its rings that seem to levitate in space and the constellations are depicted on the ground. Tourists visiting the observatory stop for a night at this huge astronomy theme park. There is a daily limit for visitors to the telescope, and electronic devices are not permitted within a five-kilometre radius, but the presence of a whole town with Wi-Fi and thousands of mobile phones on its doorstep, in spite of the protection offered by the mountains, cannot fail to create the interference that astronomers try to avoid by building their observatories in isolated places like this used to be (the same reason that thousands of people from the surrounding villages were forcibly removed). As was hoped, the FAST radio telescope led to development, but development is now threatening to create problems for FAST.

Bryan of Nevada, who said he hoped it would 'be the end of Martian-hunting season at the taxpayer's expense'. That's one reason it is China, and not the United States, that has built the first world-class radio observatory with SETI as a core scientific goal.

SETI does share some traits with religion. It is motivated by deep human desires for connection and transcendence. It concerns itself with questions about human origins, about the raw creative power of nature and about our future in this universe – and it does all this at a time when traditional religions have become unpersuasive to many. Why these aspects of SETI should count against it is unclear. Nor is it clear why Congress should find SETI unworthy of funding,

'The search for extraterrestrial intelligence does share some traits with religion. It is motivated by deep human desires for connection and transcendence. It concerns itself with questions about human origins, about the raw creative power of nature and about our future in this universe.'

given that the US government has previously been happy to spend hundreds of millions of taxpayer dollars on ambitious searches for phenomena whose existence was still in question. The expensive, decades-long missions that found black holes and gravitational waves both commenced when their targets were mere speculative possibilities. That intelligent life can evolve on a planet is not a speculative possibility, as Darwin demonstrated. Indeed, SETI might be the most intriguing scientific project suggested by Darwinism.

Even without federal funding in the United States, SETI is now in the midst of a global renaissance. Today's telescopes have brought the distant stars nearer, and in their orbits we can see planets. The next generation of observatories is now clicking on, and with them we will zoom into these planets' atmospheres. SETI researchers have been preparing for this moment. In their exile, they have become philosophers of the future. They have tried to imagine what technologies an advanced civilisation might use, and what imprints those technologies would make on the observable universe. They have figured out how to spot the chemical traces of artificial pollutants from afar. They know how to scan dense star fields for giant structures designed to shield planets from a supernova's shock waves.

In 2015 Russian billionaire Yuri Milner poured $100 million of his own cash into a new SETI programme led by scientists at UC Berkeley. The team performs more SETI observations in a single day than took place during entire years just a decade ago. In 2016 Milner sank another $100 million into an interstellar-probe mission. A beam from a giant laser array, to be built in the Chilean high desert, will wallop dozens of wafer-thin probes more than four light years to the Alpha Centauri system to get a closer look at its planets. Milner told me the probes' cameras might be able to make out individual continents. The Alpha Centauri team modelled the radiation that such a beam would send out into space and noticed striking similarities to the mysterious 'fast radio bursts' that Earth's astronomers keep detecting, which suggests the possibility that they are caused by similar giant beams, powering similar probes elsewhere in the cosmos.

Andrew Siemion, the leader of Milner's SETI team, is actively looking into this possibility. He visited the Chinese dish while it was still under construction to lay the groundwork for joint observations and to help welcome the Chinese team into a growing network of radio observatories that will cooperate on SETI research, including new facilities in Australia, New Zealand and South Africa. When I joined Siemion for overnight SETI observations at a radio observatory in West Virginia in the fall of 2016, he gushed about the Chinese dish. He said it was the world's most sensitive telescope in the part of the radio spectrum that is 'classically considered to be the most probable place for an extraterrestrial transmitter'.

Before I left for China, Siemion warned me that the roads around the observatory were difficult to navigate, but he said I'd know I was close when my phone reception went wobbly. Radio transmissions are forbidden near the dish, lest scientists there mistake stray electromagnetic radiation for a signal from the deep. Supercomputers are still sifting through billions of false positives collected during previous SETI observations, most caused by human technological interference.

My driver was on the verge of turning back when my phone reception finally began to wane. The sky had darkened in the five hours since we'd left sunny Guiyang. High winds were whipping between the *Avatar*-style mountains, making the long bamboo stalks sway like giant green feathers. A downpour of fat droplets began splattering the windshield just as I lost service for good.

*

The week before, Liu and I had visited a stargazing site of a much older vintage. In 1442, after the Ming dynasty moved China's capital to Beijing, the emperor broke ground on a new observatory near the Forbidden City. More than twelve metres high, the elegant, castle-like structure came to house China's most precious astronomical instruments.

No civilisation on Earth has a longer continuous tradition of astronomy than China, whose earliest emperors drew their political legitimacy from the sky in the form of a 'mandate of heaven'. More than 3,500 years ago China's court astronomers pressed pictograms of cosmic events into tortoiseshells and ox bones. One of these 'oracle bones' bears the earliest known record of a solar eclipse. It was likely interpreted as an omen of catastrophe, perhaps an ensuing invasion.

Liu and I sat at a black-marble table in the old observatory's stone courtyard. Centuries-old pines towered overhead, blocking the hazy sunlight that poured down through Beijing's yellow, polluted sky. Through a round red portal at the courtyard's edge, a staircase led up to a turret-like observation platform where a line of ancient astronomical devices stood, including a giant celestial globe supported by slithering bronze dragons. The starry globe was stolen in 1900, after an eight-country alliance stormed Beijing to put down the Boxer Rebellion. Troops from Germany and France flooded into the courtyard where Liu and I were sitting and made off with ten of the observatory's prized instruments.

The instruments were eventually returned, but the sting of the incident lingered. Chinese schoolchildren are still taught to think of this general period as the 'century of humiliation', the nadir of China's long fall from its Ming-dynasty peak. Only recently has China regained its geopolitical might, after opening to the world during Deng Xiaoping's 1980s reign. Deng evinced a near-religious reverence for science and technology, a sentiment that is undimmed in Chinese culture today. The country is on pace to outspend the United States on research and development this decade, but the quality of its research varies a great deal. According to a study published in *Nature* in 2010, even at China's most prestigious academic institutions, around a third of scientific papers contained plagiarised material.

It remains an open question whether Chinese science will ever catch up with that of the West without a bedrock political commitment to the free exchange of ideas. China's persecution of dissident scientists began under Mao, whose ideologues branded Einstein's theories 'counter-revolutionary'. But it did not end with him. Even in the absence of overt persecution, the country's 'great firewall' handicaps Chinese scientists, who have difficulty accessing data published abroad.

China has learned the hard way that spectacular scientific achievements confer prestige upon nations. The Celestial Kingdom looked on from the sidelines as Russia flung the first satellite and human being into space, and then again when American astronauts spiked the Stars and Stripes into the lunar crust.

China has largely focused on the applied sciences. It built the world's fastest supercomputer, spent heavily on medical research and planted a 'great green wall' of forests in its north-west as a last-ditch effort to halt the Gobi Desert's spread. Now China is bringing its immense resources to bear on the fundamental sciences. The country plans to build an atom smasher that will conjure thousands of 'god particles' out of the ether, in the same time it took CERN's Large Hadron Collider to strain out a handful. It is also eyeing Mars. In the techno-poetic idiom of the 21st century, nothing would symbolise China's rise like a high-definition shot of a Chinese astronaut setting foot on the red planet. Nothing except, perhaps, first contact.

*

At a security station fifteen kilometres from the dish, I handed my cell phone to a guard. He locked it away in a secure compartment and escorted me to a pair of metal detectors so I could demonstrate

RADIO STAR

At the end of 2020 the announcement that a radio signal potentially coming from Proxima Centauri, our closest star, had been detected over a year earlier by the Parkes Observatory in Australia sent believers in extraterrestrial intelligence into a frenzy. The signal was recorded as part of entrepreneur Yuri Milner's 'Breakthrough Listen' initiative and was dubbed BLC1 (Breakthrough Listen Candidate 1). It lasted nearly five hours and had a single frequency, 982 megahertz, with a very narrow band – all highly unusual features for a natural radio source. The frequency also varied in a way that could correspond to the Doppler effect caused by the orbital motion of Proxima Centauri b, one of the two planets in orbit around Proxima Centauri. Discovered in 2016, Proxima Centauri b is an exoplanet that is potentially temperate and rocky like the Earth (although 20 per cent larger) and located within the star's habitable zone, making it a good candidate for life even though, as some have observed, statistically speaking it would be absurd if among all the billions of galaxies, life was concentrated within this pocket-handkerchief-sized portion of the Milky Way in two neighbouring systems. Over time, explanations are found for the majority of potential extraterrestrial radio signals: they might have natural origins, such as a pulsar, or a terrestrial source (not just satellites but also aeroplanes and even microwave ovens) – as was the case, disappointingly, for BLC1, according to two later studies. There is one example, however, that has still not been linked with certainty to any source: the famous 'Wow!' signal of 1977, detected by Jerry Ehman. With his telescope aimed at a group of stars known as Chi Sagittarii, the astronomer recorded an extremely powerful signal lasting seventy-two seconds, causing him to write the word 'Wow!' on the printout in red pen.

Liu Cixin is regarded – along with Wang Jinkang and Han Song – as one of the 'three generals' of Chinese science-fiction writing. The genre is termed *chaohuan*, or the 'ultra-unreal', and takes inspiration from the hallucinatory reality of the contemporary world with an often dark view of technological development and the evolution of society. The recent science-fiction boom in China has been put down in part to realist literature's inability to keep up with the country's rapid changes. Covers of Chinese (**right**) and Japanese (**above**) editions of Liu's books.

China is starting to export its science-fiction films. *The Wandering Earth*, based on a short-story collection by Liu Cixin, has been distributed around the world by Netflix and was the second-highest-grossing non-US film in the history of cinema after *The Battle at Lake Changjin* (2021), another Chinese film.

'The collective energy of all the radio waves caught by Earth's observatories in a year is less than the kinetic energy released when a single snowflake comes softly to rest on bare soil. Collecting these ethereal signals requires technological silence.'

that I wasn't carrying any other electronics. A different guard drove me on a narrow access road to a switchback-laden stairway that climbed eight hundred steps up a mountainside, through buzzing clouds of blue dragonflies, to a platform overlooking the observatory.

Until a few months before his death in September 2017, the radio astronomer Nan Rendong was the observatory's scientific leader, and its soul. It was Nan who had made sure the new dish was customised to search for extraterrestrial intelligence. He'd been with the project since its inception in the early 1990s, when he used satellite imagery to pick out hundreds of candidate sites among the deep depressions in China's karst mountain region.

Apart from microwaves, such as those that make up the faint afterglow of the Big Bang, radio waves are the weakest form of electromagnetic radiation. The collective energy of all the radio waves caught by Earth's observatories in a year is less than the kinetic energy released when a single snowflake comes softly to rest on bare soil. Collecting these ethereal signals requires technological silence. That's why China plans to one day put a radio observatory on the dark side of the moon, a place more technologically silent than anywhere on Earth. It's why, over the course of the past century, radio observatories have sprouted, like cool white mushrooms, in the blank spots between this planet's glittering cities. And it's why Nan went looking for a dish site in the remote karst mountains. Tall, jagged and covered in subtropical vegetation, these limestone mountains rise up abruptly from the planet's crust, forming barriers that can protect an observatory's sensitive ear from wind and radio noise.

After making a shortlist of candidate locations, Nan set out to inspect them on foot. Hiking into the centre of the Dawodang depression, he found himself at the bottom of a roughly symmetrical bowl, guarded by a nearly perfect ring of green mountains, all formed by the blind processes of upheaval and erosion. More than twenty years and $180 million later, Nan positioned the dish for its inaugural observation – its 'first light', in the parlance of astronomy. He pointed it at the fading radio glow of a supernova, or 'guest star', as Chinese astronomers had called it when they recorded the unusual brightness of its initial explosion almost a thousand years earlier.

After the dish was calibrated it started scanning large sections of the sky. Andrew Siemion's SETI team worked with the Chinese to develop an instrument to piggyback on the telescope's wide sweeps, which by themselves will constitute a radical expansion of the human search for the cosmic other.

In 2017, in images taken by a telescope on the Hawaiian island of Maui, the astronomer Robert Weryk noted a dark-red object ten times as long as it was wide. It was thirty-three million kilometres away and already moving away from the sun. Once it was established that it had come from outside the solar system (more or less from the direction of Vega, a star in the Lyra constellation), the astronomers had to invent a new classification: I, for interstellar, and 1, because it was the first object of this type to have been catalogued. It is also known more evocatively as 'Oumuamua, the Hawaiian word for scout. It is now too far away to be observed. So what have we missed? What left astronomers perplexed was its trajectory. The galactic intruder moved as if it were not only obeying gravity but had its own propulsive force. Various plausible proposals were put forward – an asteroid, a comet, a cosmic iceberg – but none of them was entirely convincing – and there are those who think it was an alien artefact. In *Extraterrestrial: The First Sign of Intelligent Life Beyond Earth* (Mariner, 2021, USA / John Murray, 2022, UK), the respected professor Avi Loeb, of Harvard University, not only rigorously presented his theory but also came up with a wager, similar to Pascal's wager regarding the existence of God: 'If we dare to wager that 'Oumuamua was a piece of advanced extraterrestrial technology, we stand only to gain,' he wrote. Such a bet 'could have a transformative effect on our civilization', and he hopes that the next time an interstellar messenger arrives we will be ready to welcome it.

Siemion told me he's especially excited to survey dense star fields at the centre of the galaxy. 'It's a very interesting place for an advanced civilisation to situate itself,' he said. The sheer number of stars and the presence of a supermassive black hole make for ideal conditions 'if you want to slingshot a bunch of probes around the galaxy'. Siemion's receiver will train its sensitive algorithms on billions of wavelengths, across billions of stars, looking for a beacon.

Liu Cixin told me he doubts the dish will find one. In a dark-forest cosmos like the one he imagines, no civilisation would ever send a beacon unless it were a 'death monument', a powerful broadcast announcing the sender's impending extinction. If a civilisation were about to be invaded by another, or incinerated by

a gamma-ray burst, or killed off by some other natural cause, it might use the last of its energy reserves to beam out a dying cry to the most life-friendly planets in its vicinity.

Even if Liu is right, and the Chinese dish has no hope of detecting a beacon, it is still sensitive enough to hear a civilisation's fainter radio whispers, the ones that aren't meant to be overheard, like the aircraft-radar waves that constantly waft off Earth's surface. If civilisations are indeed silent hunters, we might be wise to hone in on this 'leakage' radiation. Many of the night sky's stars might be surrounded by faint halos of leakage, each a fading artefact of a civilisation's first blush with radio technology before it recognised the risk and turned off its detectable transmitters. Previous observatories could search only a handful of stars for this radiation.

China's dish has the sensitivity to search tens of thousands.

In Beijing I told Liu that I was holding out hope for a beacon. I told him I thought dark-forest theory was based on too narrow a reading of history. It may infer too much about the general behaviour of civilisations from specific encounters between China and the West. Liu replied, convincingly, that China's experience with the West is representative of larger patterns. Across history it is easy to find examples of expansive civilisations that used advanced technologies to bully others. 'In China's imperial history, too', he said, referring to the country's long-standing domination of its neighbours.

But even if these patterns extend back across all of recorded history, and even if they extend back to the murky epochs of prehistory, to when the Neanderthals vanished some time after first contact with modern humans, that still might not tell us much about galactic civilisations. For a civilisation that has learned to survive across cosmic timescales, humanity's entire existence would be but a single moment in a long, bright dawn. And no civilisation could last tens of millions of years without learning to live in peace internally. Human beings have already created weapons that put our entire species at risk; an advanced civilisation's weapons would likely far outstrip ours.

I told Liu that our civilisation's relative youth would suggest we're an outlier on the spectrum of civilisational behaviour, not a Platonic case to generalise from. The Milky Way has been habitable for billions of years. Anyone we make contact with will almost certainly be older, and perhaps wiser.

*

Moreover, the night sky contains no evidence that older civilisations treat expansion as a first principle. SETI researchers have looked for civilisations that shoot outwards in all directions from a single origin point, becoming an ever-growing sphere of technology, until they colonise entire galaxies. If they were consuming lots of energy, as expected, these civilisations would give off a tell-tale infrared glow, and yet we don't see any in our all-sky scans. Maybe the self-replicating machinery required to spread rapidly across 100 billion stars would be doomed by runaway coding errors. Or maybe civilisations spread unevenly throughout a galaxy, just as humans have spread unevenly across the Earth. But even a civilisation that captured a tenth of a galaxy's stars would be easy to find, and we haven't found a single one, despite having searched the nearest hundred thousand galaxies.

Some SETI researchers have wondered about stealthier modes of expansion. They have looked into the feasibility of 'Genesis probes', spacecraft that can seed a planet with microbes, or accelerate evolution on its surface, by sparking a Cambrian explosion like the one that juiced biological creativity on Earth. Some have even searched for evidence that such spacecraft might have visited this planet by looking for encoded messages in our DNA – which is, after all, the most robust informational storage medium known to science. They, too, have come up empty. The idea that civilisations expand ever outwards might be woefully anthropocentric.

Liu did not concede this point. To him, the absence of these signals is just further evidence that hunters are good at hiding. He told me that we are limited in how we think about other civilisations, 'Especially those that may last millions or billions of years,' he said. 'When we wonder why they don't use certain technologies to spread across a galaxy, we might be like spiders wondering why humans don't use webs to catch insects.' And anyway, an older civilisation that has achieved internal peace may still behave like a hunter, Liu said, in part because it would grasp the difficulty of 'understanding one another across cosmic distances'. And it would know that the stakes of a misunderstanding could be existential.

First contact would be trickier still if we encountered a postbiological artificial intelligence that had taken control of its planet. Its world view might be doubly alien. It might not feel empathy, which is not an essential feature of intelligence but instead an emotion installed by a particular evolutionary history and culture. The logic behind its actions could be beyond the powers of the human imagination. It might have transformed its

THE PASSENGER   Ross Andersen

The arrival of aliens on Earth is one of the most frequent subjects in science-fiction cinema, either in horror films such as the US–Japanese co-production *The Green Slime* (1968) and *Plan 9 from Outer Space* (1959), which has been described as 'the worst film of all time', or in the output of acclaimed directors such as Steven Spielberg and Robert Zemeckis, whose film *Contact* (1997) was based on a Carl Sagan novel.

'Cool and concave, the dish looked at one with the Earth. It was as though God had pressed a perfect round fingertip into the planet's outer crust and left behind a smooth, silver print.'

entire planet into a supercomputer, and, according to Oxford University researchers Anders Sandberg, Stuart Armstrong and Milan Circovic, it might find the current cosmos too warm for truly long-term, energy-efficient computing. It might cloak itself from observation and power down into a dreamless sleep lasting hundreds of millions of years, until such time when the universe has expanded and cooled to a temperature that allows for many more epochs of computing.

*

As I came up the last flight of steps to the observation platform, the Earth itself seemed to hum like a supercomputer, thanks to the loud, whirring chirps of the mountains' insects, all amplified by the dish's acoustics. The first thing I noticed at the top was not the observatory but the karst mountains. They were all individuals, lumpen and oddly shaped. It was as though the Mayans had built giant pyramids across hundreds of square miles, and they'd all grown distinctive deformities as they were taken over by vegetation. They stretched in every direction, all the way to the horizon, the nearer ones dark green, and the distant ones looking like blue ridges.

Amid this landscape of chaotic shapes was the spectacular structure of the dish. Five football fields wide and deep enough

to hold two bowls of rice for every human being on the planet, it was a genuine instance of the technological sublime. Its vastness reminded me of Utah's Bingham copper mine but without the air of hasty, industrial violence. Cool and concave, the dish looked at one with the Earth. It was as though God had pressed a perfect round fingertip into the planet's outer crust and left behind a smooth, silver print.

I sat up there for an hour in the rain, as dark clouds drifted across the sky, throwing warbly light on the observatory. Its thousands of aluminium-triangle panels took on a mosaic effect: some tiles turned bright silver, others pale bronze. It was strange to think that if a signal from a distant intelligence were to reach us any time soon, it would probably pour down into this metallic dimple in the planet. The radio waves would ping off the dish and into the receiver. They'd be pored over and verified. International protocols require the disclosure of first contact, but they are non-binding. Maybe China would go public with the signal but withhold its star of origin, lest a fringe group send Earth's first response. Maybe China would make the signal a state secret. Even then, one of its international partners could go rogue. Or maybe one of China's own scientists would convert the signal into light pulses and send it out beyond the great firewall to

If we really cannot find any form of life in space, we could always move things along with direct intervention, helping a planet to create the right characteristics. In reality, the history of our Earth shows that even under ideal conditions the development of complex life forms is far from simple, as the name suggests, and takes billions of years. But planets do not always have all the time in the world – in some cases their windows of habitability might last 'only' a couple of billion or even a few hundred million years, and time, as we know, flies. Only in these cases, according to some advocates of the need to send out 'Genesis probes', would it be ethically justified to help our (future) neighbours by saving them a few billion years corresponding to the evolution of single-celled organisms and taking them to the next stage. Our delivery service would initially include cyanobacteria and eukaryotes, and only at a second stage more complex heterotrophs, provided that the probe orbiting a suitable planet did not discover that it already had its own biosphere, at which point the probe would need to change target, because the aim of the Genesis project is to create life, not destroy it. No other space mission would ever have operated on such a long-term timescale: between the voyage and the evolution process on the chosen planet it could take a hundred million years or so to see the results. So would it be it worth it? The answer from Claudius Gros, systems theorist at Frankfurt's Goethe University and principal supporter of the project, is a simple one: yes, life is beautiful!

fly freely around the messy snarl of fibre-optic cables that spans our planet.

In Beijing I had asked Liu to set aside dark-forest theory for a moment. I asked him to imagine the Chinese Academy of Sciences calling to tell him it had found a signal. How would he reply to a message from a cosmic civilisation? He said that he would avoid giving a too detailed account of human history. 'It's very dark,' he said. 'It might make us appear more threatening.' In *Blindsight*, Peter Watts's 2006 novel of first contact, mere reference to the individual self is enough to get us profiled as an existential threat. I reminded Liu that distant civilisations might be able to detect atomic-bomb flashes in the atmospheres of distant planets, provided they engage in long-term monitoring of life-friendly habitats, as any advanced civilisation surely would. The decision about whether to reveal our history might not be ours to make.

Liu told me that first contact would lead to a human conflict, if not a world war. This is a popular trope in science fiction. In the 2016 Oscar-nominated film *Arrival*, the sudden appearance of an extraterrestrial intelligence inspires the formation of apocalyptic cults and nearly triggers a war between world powers anxious to gain an edge in the race to understand the aliens' messages. There is also real-world evidence for Liu's pessimism: when Orson

Cumulative daily UFO sightings (1995–2018)

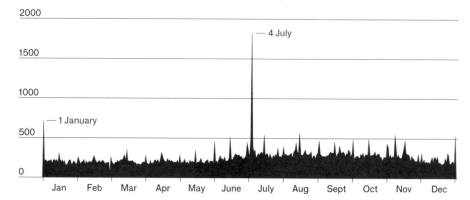

According to the US National UFO Reporting Center, a nonprofit organisation that has collected reports of Unidentified Flying Objects in the United States since 1974, sightings tend to increase on 4 July, US Independence Day. The phenomenon could be attributable to Hollywood. In 1997, the year after the film *Independence Day* was released, sightings on 4 July increased tenfold over previous years. Alcohol and other stimulants as well as fireworks might well account for the rest.

SOURCE: NATIONAL UFO REPORTING CENTER

Welles's *War of the Worlds* radio broadcast simulating an alien invasion was replayed in Ecuador in 1949, a riot broke out, resulting in the deaths of six people. 'We have fallen into conflicts over things that are much easier to solve,' Liu told me.

Even if no geopolitical strife ensued, humans would certainly experience a radical cultural transformation, as every belief system on Earth grappled with the bare fact of first contact. Buddhists would get off easy: their faith already assumes an infinite universe of untold antiquity, its every corner alive with the vibrating energies of living beings. The Hindu cosmos is similarly grand and teeming. The Koran references Allah's 'creation of the heavens and the Earth, and the living creatures that He has scattered through them'. Jews

believe that God's power has no limits, certainly none that would restrain his creative powers to this planet's cosmically small surface.

Christianity might have it tougher. There is a debate in contemporary Christian theology as to whether Christ's salvation extends to every soul that exists in the wider universe or whether the sin-tainted inhabitants of distant planets require their own divine interventions. The Vatican is especially keen to massage extraterrestrial life into its doctrine, perhaps sensing that another scientific revolution may be imminent. The shameful persecution of Galileo is still fresh in its long institutional memory.

Secular humanists won't be spared a sobering intellectual reckoning with first

Observing the sky from down here on Earth can be problematic – light pollution and the distorting effects of the atmosphere require complex adjustments – which is why astronomers prefer looking at space *from* space. For over thirty years the Hubble Space Telescope has been our best tool for exploring the universe, leading to breakthroughs such as determining its rate of expansion as well as providing us with amazing screensavers! But Hubble is getting old and is expected to become obsolete some time in the coming decade. Enter the James Webb Space Telescope (JWST), launched on 25 December 2021, eleven years late and over $9 billion over budget – still, it is the best Christmas present any astronomer could ever have wished for. Its greatly improved infrared resolution and sensitivity allows it to view objects too old, too distant and too faint for Hubble and allows for a broad range of investigations, such as observations of first stars and the formation of the earliest galaxies as well as the detailed atmospheric characterisation of potentially habitable exoplanets. Its technical specifications are mind-numbing, including a gold-coated beryllium mirror, 6.5 metres in diameter, made up of eighteen separate hexagonal mirrors which all unfolded in unison after the launch. Another advantage over Hubble will be its position much deeper into space, at what is called the second Lagrange point (L2), a perfect parking spot where it can use the bare minimum of fuel to orbit thanks to its alignment with the sun and the Earth. As its first observations are transmitted down to us, NASA and other agencies are already looking ahead to the next big project, the Nancy Grace Roman Space Telescope, scheduled for 2027, which will complement the JWST's work while trying to answer some basic questions about dark energy.

contact. Copernicus removed Earth from the centre of the universe, and Darwin yanked humans down into the muck with the rest of the animal kingdom. But even within this framework, we human beings have continued to regard ourselves as nature's pinnacle. We have continued treating 'lower' creatures with great cruelty. We have marvelled that existence itself was authored in such a way as to generate, from the simplest materials and axioms, beings like us. We have flattered ourselves that we are, in the words of Carl Sagan, 'the universe's way of knowing itself'. These are secular ways of saying we are made in the image of God.

We may be humbled to one day find ourselves joined, across the distance of stars, to a more ancient web of minds, fellow travellers in the long journey of time. We may receive from them an education in the real history of civilisations – young, old and extinct. We may be introduced to galactic-scale artworks, borne of million-year traditions. We may be asked to participate in scientific observations that can be carried out only by multiple civilisations, separated by hundreds of light years. Observations of this scope may disclose aspects of nature that we cannot now fathom. We may come to know a new metaphysics. If we're lucky, we will come to know a new ethics. We'll emerge from our existential shock feeling newly alive to our shared humanity. The first light to reach us in this dark forest may illuminate our home world, too. 🖋

The Cosmic Comedy

FRANK WESTERMAN

Translated by Laryssa Aijal

A monument near the railway station in Plesetsk, four kilometres from the cosmodrome in Arkhangelsk Oblast, eight hundred kilometres north of Moscow and around two hundred kilometres south of the city of Arkhangelsk.

The conquest of space was driven by the rivalry between the Cold War superpowers, and yet space is also where we like to project our utopias of unity and peaceful coexistence. Far from Earth humanity displays the best version of itself. But if we are unable to work together down here, what makes us think we will be able to up there?

FRANK WESTERMAN is an award-winning Dutch author and journalist – he reported from Srebrenica during the Bosnian civil war and was Moscow correspondent for NRC Handelsblad for five years – who writes on the themes of racism, culture, identity and power. His work has been translated into more than fifteen languages, including four titles in English, *Ararat: In Search of the Mystical Mountain*, *Engineers of the Soul: The Grandiose Propaganda of Stalin's Russia*, *Brother Mendel's Perfect Horse: Man and Beast in an Age of Human Warfare* and 2022's *We Hominids: An Anthropological Detective Story* (Head of Zeus). His research into and reflections on our changing perceptions on heaven and hell in times of space travel – the source material for the essay here – are discussed in his book *De kosmische komedie* (published in the Netherlands in 2021).

RAFFAELE PETRALLA is a Rome-based photographer and filmmaker, whose work focuses on social and anthropological issues. He has received many international awards for his photography, and his pictures have been published in numerous high-profile publications across the world.

For the photographs reproduced here, Petralla travelled to the edge of the Arctic Circle and the border security zone in the Mezensky District of northern Russia. The inhabitants there build sledges, boats and other objects out of material salvaged from the spent rockets used to launch satellites from the Plesetsk Cosmodrome, which became a very important site after the collapse of the USSR, as the main Soviet-era base at Baikonur is now located in independent Kazakhstan.

I

The world's first spacewalker is a miner's son. As his father lowers himself into the subterranean shafts, Alexei Leonov rises up through the thin atmosphere. Unlike the ascension of Christ, his journey into space is well documented. There is film footage of it.

An R-7 rocket lifts cosmonaut Leonov into orbit. Weightlessly he whizzes around the planet. Through the porthole of his spacecraft he witnesses sixteen sunsets and sunrises in a single day. The Siberian has been hurled out of the atmosphere on a secret mission. On 18 March 1965 at 11.30 a.m. (Moscow time) he receives orders to enter the airlock. The voice commanding him emanates from a pine forest somewhere on Earth. The code name of his flight controller is Zarya. Dawn.

Alexei, thirty years old, manoeuvres himself out of his metal capsule and into the 'accordion' airlock, an inflatable cylinder with a hatch at the far end. Once inside this rubber appendage he must acclimatise for fifty minutes to prevent nitrogen bubbles from boiling in his blood, as can happen to divers who surface too rapidly.

His pulse rises from eighty-six to ninety-five, and upon the command to open the hatch it shoots up to 150. Overheating and soaked with sweat, an oxygen tank on his back, Alexei squeezes himself out of the airlock – and his white figure steps into the void. Without leaving the side of his mothership, he floats freely through the universe connected only by an umbilical cord.

The coastline of North Africa passes by far below him, followed swiftly by those of Turkey and Crimea. The Caucasus looks partly overcast.

Cheering can be heard from the sublunary world. 'What you have accomplished is beyond the boldest imagination,' the Kremlin declares. Comrade Leonov has 'courageously opened the door to the universe'.

Generations of earthlings have been eagerly awaiting this breakthrough. More than simply a Soviet citizen sporting the Cyrillic letters CCCP on his helmet, Alexei is an ambassador for all of us. The 'invasion' of space by a member of the world's population transcends the battle of prestige between East and West. From the movements of his limbs, somewhere between treading water and swimming, it seems Alexei is engaged in a struggle.

Where George slew the dragon, cosmonaut Leonov duels with God. The stakes: from this moment on, who will rule in heaven?

Alexei has barely stepped on to the field of battle when he is dealt a heavy blow. The heavenly throne will not be ascended without a fight. In the vacuum of space Alexei's suit inflates more than anticipated. His hands no longer reach into his gloves; his feet no longer reach into his boots. Outside the airlock the miner's son swells up like the Michelin Man.

Below him the Black Sea drifts by in a westerly direction. It will be dark fifteen minutes from now. At ten minutes in, he prepares to crawl back into the belly of the mothership. But he can no longer fit through the hatch. Alexei is a wrestling champion, nicknamed 'the Siberian Bear', but no matter how much force he puts into it, his suit is too big.

Radio Moscow does not wait. The state broadcaster wastes no time in celebrating his spacewalk as an accomplished feat of arms – with Alexei's four-year-old daughter Viktoria live in the studio.

'Where is your father now?'

'Daddy's floating in space,' listeners hear her say.

II

Humanity has introduced itself to the rest of the universe in a kind of celestial advertisement through the gold phonograph discs on board *Voyagers 1* and *2*.

In 1977 NASA's curious duo of explorers were sent to Jupiter and Saturn to investigate. They were equipped with umbrella-shaped antennas to keep in touch with home during their grand tour. As they raced past Jupiter (in 1979) and Saturn (1980, *Voyager 1*; 1981, *Voyager 2*), they performed some daredevil stunts to get the best photographs. Never before have the sulphuric volcanoes of Io – first discovered by Galileo – and the rings of Saturn been seen in such detail. There are a mere three centuries separating the moment a telescope revealed their existence and this first encounter, no more than the blink of an eye on the astronomical timescale. *Voyager 2* has also flown past Uranus (1986) and Neptune (1989). One planet's gravitational field alters the vessel's trajectory to steer it towards the next, and this, swinging their way through the solar system like monkeys in the treetops, is how the *Voyagers* plot their course.

A return to Earth is not an option; their mission is open ended. Having gone beyond the furthest planet, *Voyager 1* (in 1990) turned around once more to take a farewell photo. We are in it, far away and therefore rather small. Our globe, seen from a distance of six billion kilometres, is an insignificant little dot in the cosmos that, with a little imagination, could be described as blue. It promptly earned the Earth a new nickname: the Pale Blue Dot. Those who call her by that name generally lower their voices and share how the sight humbles them.

Look, *Voyager* winks at us, you're all making a fuss over nothing. Or is it just a matter of time, and will future generations in the 22nd century look upon this dot with pride? 'Look, that's where we came from.'

The cameras on both probes have now been switched off to save energy. Thirty-five years after its launch, in 2012 *Voyager 1* left the solar system followed by *Voyager 2* in 2018. Ever since then they've been racing through interstellar space. This is exploration in the most adventurous sense of the word. In three hundred years' time they will traverse the Oort cloud, a belt of comets and cosmic debris discovered by the Dutch astronomer Jan Oort. Then we will wait for the passing of a star. The first to present itself will be Gliese 445 in the Camelopardalis (giraffe) constellation, which *Voyager 1* will pass at a distance of around 1.6 light years in about forty thousand years' time.

Although the chances are slim, it is not impossible that one day these wiry objects and their unfolded umbrellas might be detected and taken to be malevolent invading UFOs that have ignored all warnings to stop. Just imagine if an alien civilisation managed to catch one. Wouldn't it be a shame if there was no information about who sent it? With this in mind, the Golden Records were cut. These contain information on who we are and where we live. The discs are affixed to the exteriors of both *Voyagers* in clear view. They are humankind's business cards, round and shiny, to be offered to any non-human who finds them.

These LPs were designed to survive for at least a billion years. Other than the name of their producer (NASA), they feature an etching of the position of our sun with respect to fourteen pulsars, neutron stars that emit radiation at distinctive intervals and act like light-houses. Thus, whoever pinpoints the star that we call the sun will be able to tell from the accompanying pictogram that the Earth is the third satellite.

While a stylus has been supplied, a record player has not. Aliens smart enough to intercept a spaceship without damaging it, so the thinking goes, will be smart enough to unlock the data stored in the grooves. Just suppose that, somewhere far away, one of these Golden Records is being played. It is hard to imagine what effect that would have, but the sound of an opera being sung in the middle of a jungle might come close.

What is it that we are offering these strangers?

The title, *The Sounds of Planet Earth*, doesn't reveal much. Nor will the contents be easy to decipher. The message we wish to convey has been wrapped up in ninety minutes of audio and 115 digital images – cryptograms of an inscrutability that we normally attribute only to God.

It is the intention that counts, and the intention is good. Before we present ourselves to the alien life form, we offer an olive branch.

Shalom. Paz. Peace.

Expressions of peace and goodwill can be heard in fifty-five different tongues, including Esperanto, the language created to foster world peace:

Ni strebas vivi en paco kun la popoloj de la tuta mondo, de la tuta kosmo.

We strive to live in peace with the peoples of the whole world, of the whole cosmos.

After this volley of salutations we hear the whispering of the wind, the breaking of waves, birdsong and underwater whale sounds, after which we, those who sent the *Voyagers*, demonstrate just how far we are above nature: Bach, Beethoven, Mozart and Stravinsky (*The Rite of Spring*) represent

'The *Voyagers* will paint a sunny picture of our time on Earth, an autobiography redacted to such a degree that it seems to have been drafted deliberately to fool not only alien beings but, primarily, ourselves.'

classical music, Louis Armstrong plays some blues and Chuck Berry some rock 'n' roll ('Johnny B. Goode'). For the sake of inclusivity, a selection of 'world music' has been included, ranging from Senegalese percussion and Georgian polyphonic singing to Javanese gamelan.

On these gold discs the human race is laying out its beads and mirrors. During the ninety-minute revue we put our best foot forward. On the off-chance our message in a bottle is ever found and deciphered on some distant planet, we are making sure we look sharp. We have taken a self-portrait of how we like to see ourselves – as virtuous creatures, happy, peaceful – and sent it off into the Milky Way. That is how we hope to introduce ourselves to the cosmos. There are a few things missing from our best-of collection, though: poverty and hunger, disease and pain, jealousy and greed, waste and pollution, murder and manslaughter. The pictures we have sent by way of introduction show people eating and drinking, not urinating or defecating. There is a smile, not a tear.

During that same year of 1977 Red Army Faction terrorists combat capitalism, Bokassa crowns himself emperor in what had been the Central African Republic, Charter 77 human-rights activists are arrested in Prague and a war between Somalia and Ethiopia breaks out in the Ogaden. Not a hint or even an inkling of any of this is to be found on these discs sent heavenwards. Human nature's demonic, destructive side has been carefully glossed over.

Should we be wiped out, the *Voyagers* will be our last testament. They will paint a sunny picture of our time on Earth, an autobiography redacted to such a degree that it seems to have been drafted deliberately to fool not only alien beings but, primarily, ourselves.

III

I bought a copy of the record (the deluxe edition, on vinyl) and played it at home. Track 1, 'Greeting from Kurt Waldheim, Secretary-General of the United Nations', is forty-four seconds long: 'I send greetings on behalf of the people of our planet,' says Kurt Waldheim. In a measured voice and on behalf of all humankind, the Austrian speaks to our fellow inhabitants of the universe: 'We step out of our solar system into the universe seeking only peace and friendship, to teach if we are called upon, to be taught if we are fortunate.'

But this speaker is not who he seems. At the time of recording this message of peace in 1977 Dr Kurt Waldheim had not yet been exposed as an ex-army officer who wore a uniform of the Nazi regime, whose unit sent forty thousand Thessaloniki Jews to their deaths. Until the late 1980s Waldheim would deny having been awarded the silver medal (with oak leaves) of the Order of King Zvonimir in Hitler's puppet state of

Above: Pavel Popov, 46, started looking for rocket remains during the perestroika era in the village of Dolgoshechlye, within the border security zone in the Mezensky District. 'When they fall very close to the village you can see the windows shaking and hear a loud noise. After the first boom, you usually hear more, which are the other three rockets.'
Below: Rocket parts in the yard of a metalworker who collects and recycles them. The outer sections are fully recycled and used to make boats and sledges.

Above: A fragment of a rocket that fell to Earth emerges from the February snow covering the tundra. Pavel Popov digs to extract and recycle it; his shovel is itself made from part of a rocket.
Below: Cousins Oleg and Aleksei Titov resting on their boat. During the months of the year when Arkhangelsk Oblast is free of snow, they ply the river transporting essential supplies from village to village.

> 'I distrust the superlatives with which NASA garlands the International Space Station. Is this peace lab not surrounded by thousands of spy satellites that keep us all caught in a web of suspicion?'

Croatia in 1942 for his part in the hunting down of partisans in Bosnia. Kurt Waldheim had served under General Alexander Löhr, who was convicted of crimes against humanity and hanged in Yugoslavia in 1947. Although it was never proved that Lieutenant Waldheim himself committed torture or murder, he did serve under a commander who would order pamphlets to be dropped behind enemy lines that read: 'Kill the Jews, come on over'.

IV

Other than the two business cards, humankind has put up another signpost in space. This colossus, a collection of interconnected modules, has been operational since the year 2000. If you include its solar panels, the construction spans the length and width of a football field. The goal of the International Space Station (ISS) is to demonstrate that we are able to rise above earthly feuds by inhabiting space together in harmony. No expense has been spared to realise this ambition. To date more than $100 billion has been spent on building this castle in the air. The ISS is the most expensive structure ever built.

Way up yonder there is constant toing and froing of astronauts and cosmonauts – more than two hundred, both men and women, from nineteen different countries so far. To honour the twenty years of continuous habitation, NASA put out a video in August 2020. A female American voice describes the multicultural space house as a 'bridge between and above all nations'. The spaceship, heavier than the heaviest oil tanker, is 'a shining example of international peace and collaboration'. Synthesisers sound. The people of the Earth haven't just raised a flag out there beyond the atmosphere but 'a symbol of humanity at its best'.

Without any sense of pride, I, as an earthling, do not have any problems with this floating show home; what does embarrass me, however, is the red Tesla (driven by a dummy known as Starman) that Elon Musk sent into orbit around the sun in February 2018 as an extension of his own ego. This does not alter the fact, though, that I distrust the superlatives with which NASA garlands the ISS. Is this peace lab not surrounded by thousands of spy satellites that keep us all caught in a web of suspicion?

Over images of smiling Russians and Americans swimming into each other's arms, the script reads: 'What started as an accord between two former rivals became a beacon of opportunity for the rest of the world.' Is this simply PR, or should I defer judgement and wait to see if perhaps peaceful coexistence in space *is* possible? Who knows, maybe it will help the human race to start over elsewhere with the slate wiped clean.

V

In November 1998 I attended the launch of the first habitation module of the ISS. It is called *Zarya*. Dawn. On the journey from Moscow to Baikonur and back – all inclusive for $1,500 per invitee – we would witness a series of culture clashes. The first occurred at the reception desk of the Hotel National near Red Square: the NASA delegation was baffled by the excessive luxury of the marble-tiled interior. How could such splendour have survived communism? Not until the next morning in Kazakhstan, after a tiring overnight flight, did the surroundings start to shape up to expectations. Although the word 'cosmodrome' may sound magical, everything looked dreary: the checkpoints along the way, stray camels dropping tufts of loose fur, sun-bleached curtains in the minibuses that ferried us to the launch site. 'Got it off some communist,' an American fellow reporter replied when I asked her about the Sputnik badge on her scarf.

The only NASA representative we were allowed to quote complained about how the Russians would not abandon the *Mir* space station. They were clinging on to the last vestige of Soviet pride, but this ate into the already limited resources for building the new communal space quarters.

We passed the tracks along which a Proton carrier rocket had been transported at walking pace from its hangar to the launch pad. There was a wall which bore an inscription in Russian: 'Russia was, is and will always be a space superpower'. Our hosts pointed towards a stone-built house with a well-manicured garden. This was where Comrade Gagarin had prepared himself for his cosmic stunt of 12 April 1961.

The rest of Kazakhstan looked godforsaken and desolate. It was either too

Although it is the world's oldest and largest spaceport – and, according to many, also the most illustrious – its existence remained somewhat mysterious for years. During the space race the Soviet Union tried to put its rivals off the scent by giving the cosmodrome (which covers nearly five times the area of the whole of Greater London) in the arid Kazakh steppe the name of a mining city that was actually 320 kilometres to the east of the site that launched the first man into space. In 1994, following the uncertainty of the early 1990s, Russia reached an agreement with Kazakhstan, which had been independent since 1991, to lease the site until 2050. Russia pays $115 million a year for the lease, although it receives over $70 million from the USA for every passenger heading up to the International Space Station aboard a Soyuz spacecraft. The exclave is governed by Russian law, even though the mayor is appointed by mutual agreement with Kazakhstan, which also takes care of law enforcement. In the golden era the town that grew up by the base to house technicians and staff was called Leninsk (it has only officially been Baikonur since 1994); its population reached 120,000 but is now less than a third of that. Russians account for the lion's share of those who have left and now make up only 35 per cent of the population. The base's decline is also down to internal competition from the new Vostochny Cosmodrome, established by Vladimir Putin and inaugurated in 2016 in Russia's far east, which will gradually take over from Baikonur ahead of its planned closure in 2030.

hot or too cold. In the middle of the steppe we got out at a concrete amphitheatre. In the distance, on launch pad A-92, was the Proton with the *Dawn* module balanced on its nose. NBC and ABC news teams were not happy with their assigned seats on the stand. 'Is there no countdown?' I heard a cameraman say. Counting down brings bad luck: according to the Russians *nol* (zero) is too reminiscent of the sense of disappointment that follows a bubble bursting. Urinating against the left front wheel of the bus that drives the cosmonauts to the launch tower, however, will guarantee a victorious mission. Yuri Gagarin had done this, and, ever since, it has been the customary last stop before leaving Earth.

Suddenly the Russians called '*Pusk!*'

The ABC crew seemed taken aback: 1,100 metres away the sand of the steppe rose in a fury. On the launch pad six roaring orange orbs fused into a single flame. The Proton ascended like an inverted white candle above the dust cloud, and after 126 seconds graciously discarded the spent fuel tanks and, describing an arc in the sky, disappeared from view.

The Americans started to clap; the Russians raised a gloomy *Hoorah!* Waiters, appearing out of nowhere, invited us to stand in a circle in the paved square. They served sweet champagne and vodka. Russian officers in fur hats made one toast after another: 'to the friendship of nations', 'to world peace', 'to the ladies present' and, above all, 'to US–Russian cooperation'. Despite the rhythmic emptying of glasses, the party never really took off. There was still too much resentment and mistrust bubbling under the surface. Even though the Cold War was behind us, the matter of who had lost and who had won hung in the air like electricity above the dying cosmodrome.

When raising this with the officers they would hint at issues other than wounded pride. They were sceptical.

About what? About the idea of a new utopia in space.

Why? Because they had personally experienced the Soviet experiment.

The vision of the phoenix that we – in the words of NASA's speechwriter – had just witnessed rising from the ashes came up against a wall of critical consideration.

'LAUNCH SUCCESSFUL, RELATIONS NOT YET', ran the newspaper headline above my Baikonur report.

The fact that the Americans had originally suggested naming the project Alpha had been an insult to the Russians. In that they heard a denial of their former glory, as if the Soviet Union had not been the world's first space power. Who had launched the first artificial satellite, the first dog, the first human, the first woman and the first to spacewalk? The Russians had only just been able to pull the name *Zarya* out of the fire. At the root of this code name was their lead engineer, Sergei Korolev. He was the Soviet state secret who had guided Yuri Gagarin (*Kedr*, Cedar), Valentina Tereshkova (*Chaika,* Seagull) and Alexei Leonov (*Almaz*, Diamond) through space by radio.

Every speech held together by 'dreams', 'new beginnings' and 'the future' would fling its Russian audience back to the time of great expectations. Whether they had believed in it or not, the nearly three hundred million citizens of the Soviet Union had been test subjects in the largest utopian project in history – on one-sixth of the inhabited area of the Earth.

VI

Out of a sense of nostalgia, cosmonauts in training still climb the stairs up to Konstantin Tsiolkovsky's old study. From

Animals began exploring space much earlier than humans, albeit not through any wish of their own. The most famous was Laika, the mongrel from the streets of Moscow who went into orbit on 3 November 1957, making her the first creature to orbit the Earth. She was followed three years later by two more little dogs, Belka and Strelka, who travelled on Sputnik 5 for more than twenty-four hours before coming back down to Earth, the first animals to return from space safe and sound. Other space travellers have included chimpanzees (Ham, on board the Mercury-Redstone 2 rocket in 1961), cats (Félicette, the only cat to have survived a space voyage), rats, tortoises, chickens, jellyfish … and spiders: in 1973 Anita and Arabella were sent up to the US space station *Skylab*. The idea of sending spiders into space to study web construction in microgravity came from seventeen-year-old Judith Miles as part of an initiative that conducted twenty-five experiments devised by students during the Skylab 3 mission. The two cross spiders selected were served houseflies before the launch that sent them off for two months to spin webs where no webs had been spun before. Spiders use their own weight to coordinate their weaving, so it was interesting to test the effects of a change in gravity. A little shy at first, the two spiders finally set to work. The webs were thinner than those spun on Earth but otherwise very similar to the 'originals'. Neither Anita nor Arabella survived the voyage, but the experiment revealed a lot about the adaptation of motor responses in space.

the training facility at Zvyozdny Gorodok (Star City) it is a three-hour drive to his log house in Kaluga on the Oka River. Tsiolkovsky is the grandfather – nay, patriarch – of space travel, the genius too poor to pay for schooling. His 1903 'rocket equation' – $v = v_1 ln(1 + m_2/m_1)$ – remains, to this day, the foundation of all rocket science. Tsiolkovsky's most famous quote, dating from 1911, is: 'Earth is the cradle of humanity, but one cannot remain in the cradle for ever.' He predicts that in the 20th century humankind will sail the cosmos like it sails the seas. The almost deaf Tsiolkovsky uses a gramophone horn as a hearing aid. In Christ he sees a humanist, and in Jesus' statement 'My Father's house has many rooms' (John 14:2) he hears a reference to the many civilisations in the universe. Already during his lifetime he is dubbed the Russian Leonardo da Vinci. As a successor to the space rocket he draws a

THE PASSENGER   Frank Westerman

'space elevator', a floating station orbiting the Earth and attached to a tether that has cable cars travelling up and down.

'To save rocket fuel,' he writes in *Dreams of the Earth and Sky.*

Tsiolkovsky believed that engineering's flywheel effect would soar so high that humanity would no longer have to fear its own extinction. Will our sun die one day? Then we will move to another solar system and travel onwards from there. Our earthly worries, he proposes, we will leave behind on Earth.

The pilgrimage to Tsiolkovsky's attic is part of cosmonaut training. In silence, aspirants examine his space equation, his drawings for the space elevator and the first edition of his 1893 science-fiction novel *Na Lune* (*On the Moon*).

'First, inevitably, the idea, the fantasy, the fairy tale,' Tsiolkosvky wrote. 'Then, scientific calculation. Ultimately, fulfilment crowns the dream.'

What makes the thought of this image particularly poignant is the fairy tale's unhappy ending. The Soviet people were given a helping of space flight that could make them proud, a sort of Olympics in space, but one without any joy. In the early days the possibilities had seemed endless and splendid. In the fervid aftermath of the Russian Revolution, artists (*liriki*) and engineers (*fisiki*) were encouraged to grow wings and fly. In the words of the poet Mayakovsky, humankind was 'about to shake off the yoke of gravity'; on the updraughts of the revolution we would elevate ourselves to become a better version of who we are.

The prose of the future, the writer Yevgeny Zamyatin said in 1919, would from now on need 'to possess the epic presence of an interplanetary journey through the frozen vacuum of the cosmos'. The starry sky became the page on which this dreamed-of utopia could be depicted. In the novel *Aelita* (1923), which was also adapted for the screen, Alexei Tolstoy had the red flag flying on Mars. The hero of the tale – in whom the audience will recognise the genial Tsiolkovsky – is a man who builds rockets with no more than a primary-school diploma. No matter how well *Aelita* was received, both book and film fell out of favour when Stalin suddenly decided he'd had enough of utopian projections on faraway planets. The ideal state was being built in the here and now for everyone to see; heaven on Earth was being built within the borders of the USSR. After the playful 1920s, Soviet artists had to knuckle down and start to work to the five-year plans.

Alexei Tolstoy stopped writing science fiction. The futurist Mayakovsky put a bullet through his heart in 1930 (because of love but also because his star was starting to fade). Yevgeny Zamyatin narrowly managed to escape to Paris in 1931. Because of its powers of prophecy, his 1921 dystopian novel *We* had been labelled subversive. Page after page its readers were drawn, just like rocket engineer D-503, into the clutches of the One State. The unsettling story is set around the docks where the *Integral* is being built, a gleaming spaceship designed to spread the One State's totalitarian ideology to new solar systems.

We is top of the list in the huge Soviet index of forbidden literature.

VII

By the time the comrades actually started their victory march outside the atmosphere, humour and self-mockery had been banished from the Soviet Union. Fear trapped laughter, censorship any criticism. Cosmonaut Alexei Leonov was required to keep silent about the near-fatal events during his spacewalk until the glasnost era of the 1980s. Looking back in 1997 he described the *Voskhod* capsule as 'the most dangerous spaceship ever to have been launched'. Because of the ballooning of his suit he could no longer squeeze through the hatch of *Voskhod 2*. Hundreds of kilometres above the Caucasus he made the decision, without consulting flight controller Zarya, to open a valve to let some air escape. 'I had no other choice,' he said apologetically. At the risk of succumbing to the bends from decompression, he lowered the internal pressure 'by about half'. As his suit started to deflate he could feel his skin beginning to tingle. It made his head swim, but before he fainted he managed, with the last reserves of his energy, to get back inside the airlock.

Apart from the swelling of his suit, it transpired that the braking rockets, required during re-entry, hadn't worked properly. Salvage helicopters were on standby in Kazakhstan, but the capsule had veered more than a thousand kilometres off course and, spinning beneath its parachute, had crashed down in a snow-covered forest in the Urals. Here Leonov waited for help for one more day and night.

'Of course, this news was never allowed to come out,' he said. 'The ideology required that everything produced by the Soviet Union was superior, never failed, simply could not fail.'

Leonov's working life, from jet pilot to cosmonaut, had been dominated by the Cold War. During the pioneering years of space travel there was no place for curiosity about the expanse. This was not about science but about power, a display

The Apollo programme took humans to the moon, and yet its most lasting legacy could well have been a byproduct: the image of our planet hanging in space. In 1968 the Apollo 8 astronaut William Anders took a photograph of the Earth, partially in shadow, rising over the surface of the moon. The image, known as *Earthrise*, inspired the nascent environmental movement: shortly afterwards Friends of the Earth and Greenpeace were founded and the first Earth Day was held. Then came *The Blue Marble*, one of the most reproduced images in history: a view of the Earth taken from Apollo 17 in 1972, with the Antarctic ice cap centre stage. That was the final Apollo mission, but NASA realised it had found a new mandate. In 1990 Carl Sagan suggested turning the camera on *Voyager 1* around to photograph Earth as a distant, barely visible 'pale blue dot'. The year before, NASA had already formalised its Mission to Planet Earth programme (now known as NASA Earth Science), with the aim of using its probes to monitor the biosphere from space. In the three decades since then NASA and other partners, including ESA, have amassed a continually expanding library of images that show the changes on Earth from land use, human activities, meteorological phenomena and the climate. As William Anders – the photographer of *Earthrise* – once said, 'We came all this way to explore the moon, and the most important thing is that we discovered the Earth.'

You might ask why anyone would fire a missile at their own satellite, but that is exactly what the space powers have been doing: China in 2007, India in 2019 and Russia in 2021. They are meant as warnings to anyone taking notice: if we can destroy one of *our* satellites, we can do the same to *yours*. The USA undertook several such tests during the Cold War – so anti-satellite weapons are nothing new – but in 2022 announced a unilateral ban: the explosions they cause are just way too dangerous, as they increase the amount of orbiting debris enormously, putting other satellites and space infrastructure at risk. New methods are being tested instead – not just projectiles but lasers and cyber attacks. Modern armies depend more and more on satellite communications, to the extent that in 2019 the USA created a new branch of its armed forces, the US Space Force, responsible for all operations in space and cyberspace. In this context, it is easy to imagine that in any possible future war between world powers the first shots will be fired in low Earth orbit. Attempts are being made at the UN to stop this new arms race, but the negotiations are bogged down in strategic and legal issues. In the meantime, the USA, which still has a significant advantage over everyone else, is increasing its capacity to build and launch new, smaller and, crucially, cheaper satellites with the help of private companies, so that anything shot down by its enemies can be rapidly replaced. The shroud of the dark side is falling. Begun the satellite wars have.

of dominance of the most primitive kind, only high tech. As a 23-year-old fighter pilot in 1957, Alexei Leonov had made a number of reconnaissance flights over East Germany. From the cockpit of his MiG-15 he could see the F-100s shooting by on the western side of the Iron Curtain, manned by pilots who, like him, would later be a part of the big mock duel in space. The competition was fierce. 'To us, the astronauts were our adversaries. We knew nothing of each other except about each other's achievements. I did wonder once, didn't we want to know more about one another?'

In the first half of the 1960s communism is way ahead of capitalism in the race into the high ground of the heavens. Space is transformed into a battleground. Astronauts and cosmonauts, helmeted and harnessed, are fighting their mock duel. Who embodies evil, who embodies good?

Time and time again the Soviets surprise the world with their firsts and their achievements. Alexei Leonov's circus act in March 1965 is yet another psychological blow to the Americans.

Back on Earth, Comrade Leonov states that he has seen the heavens, has had a good look around, but has not seen a god anywhere. '*Boga nyet*,' he says like his fellow cosmonauts. 'There is no god.' After a parade in Red Square he is awarded not one but two Orders of Lenin. The bald Siberian with the long sideburns is honoured with the title Hero of Socialist Labour. Behind the scenes, Leonov is preparing for a moon orbit, which would be followed by a moon landing. John F. Kennedy has given NASA a deadline to put a man on the moon before 1970. The Soviets accelerate their Luna programme, but with the death of Sergei 'Zarya' Korolev in January 1966, the comrades fall behind and see their hopes evaporate.

Soviet subject Alexei Leonov became a pawn, pushed strategically around the map of the world by the long arm of Moscow. 'There is practically no country on Earth where I have not signed an autograph,' he said with a hearty smile. The world's first spacewalker met Charles de Gaulle, King Juan Carlos, Pablo Picasso. He became an honorary citizen of Perm, Kaluga, Kemerovo, Nalchik, Kaliningrad, Vladimir, Arkalyk, Karaganda, Termez but also of Atlanta, New York, Los Angeles, San Antonio, Salt Lake City, Nashville and Chicago.

During Leonov's lifetime hostility slowly turned to friendship. The Soviet and American spacemen first met face to face during a convention in Athens in 1965. 'We had agreed to meet in the lobby of the hotel, but the Americans did not show up because they did not want any press involved,' Leonov recalled. 'So we went to their hotel room that evening. How did we communicate? I truly don't know. There was no interpreter, but we had brought vodka, and they had brought whiskey.'

The time of peaceful coexistence in the 1970s saw a new mood develop. The Soviet Union nominated Leonov for the first international rendezvous in space: the docking of an Apollo module with a Soyuz at 225 kilometres above the Atlantic Ocean. In a spirit of openness, the USSR broadcast the Soyuz launch from Baikonur live on television, which was followed by the Apollo launch from Cape Canaveral seven and a half hours later. On 17 July 1975 the two spacecraft slowly floated towards each other. After a few minutes of manoeuvring, the specially designed docking module joined the two craft. By opening the hatches from the inside the crews were able to visit each other. Their commanders (Leonov for the Soviets) traded a small hammer-and-sickle flag for the Stars and Stripes as a gesture of peace.

Unlike many of his fellow cosmonauts, Leonov renounced his faith in communism. Ever since the workers' paradise had begun to tear itself apart from within, Alexei Leonov had applied himself to painting. This was something he had always done (oil paintings on canvas, cosmic impressions), but in the 1990s church towers suddenly started to make their appearance on his canvases: golden onion-shaped domes, which reflected light like rockets on a launch pad.

I asked the world's first spacewalker if he had started to believe in God.

'No,' he said, 'but there is more between heaven and Earth than we humans know. I am certain of it.'

After the fall of the Soviet Union he started to advertise Omega watches. His U-turn was so radical that he was even

Reaching for the Stars: Tales of Lesser-known Space Programmes

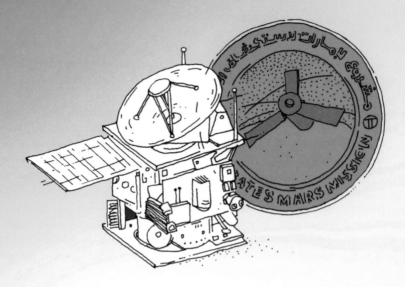

United Arab Emirates

The United Arab Emirates Space Agency (UAESA) is a new kid on the block. Established in 2014, it set itself the ambitious target of celebrating the Emirates' fiftieth anniversary, on 22 December 2021, by sending a probe to Mars as part of the first interplanetary mission led by an Arab nation. As it usually takes around ten years to organise a project of this nature, they had to move fast. After establishing the scientific objective (analysing the Martian atmosphere), work began on the technology needed to make it possible. The whole mission was put together by an international team using Zoom and other online collaboration tools. *Hope*, the first of the three probes sent to Mars in July 2020 (followed by China's *Tianwen-1* and the US rover *Perseverance*), was launched on 19

July from the Tanegashima Space Center on a Japanese rocket and successfully went into Mars orbit on 9 February 2021, making the UAESA the fifth space agency to reach the red planet. *Hope* will remain active for an entire Martian year and will study its daily and seasonal meteorological cycles, meteorological events in the lower atmosphere, such as sandstorms, and the way the climate varies across the planet's different regions. It will also try to discover why Mars is releasing hydrogen and oxygen into space and what might have been the reasons for the drastic changes to its climate. It is worth mentioning the key role played by women in the team of Emirati engineers working on the mission alongside foreign research institutes: the minister for advanced technologies, Sarah Al Amiri, was deputy project manager and lead science investigator, and women accounted for 80 per cent of the team.

Nigeria

The National Space and Research Development Agency (NASRDA), one of the leading African space agencies, was established in 1999. Nigeria had recently returned to democracy after a period of military rule between 1993 and 1998, and the timing seemed a reaction to the lack of scientific research under the military regime. *NigeriaSat-1*, the country's first satellite, was launched in 2003, and since then Nigeria has launched five more, three of which are still operational. The programme has so far only succeeded in building its own satellites, but its objectives include putting a Nigerian astronaut into space and creating a vehicle to launch satellites from a space port on Nigerian soil. Space forms an integral part of the technological independence programme that aims to encourage socioeconomic development in a country where there is still little investment in scientific education. The hope is that future Earth observation technologies will help to combat climate change and terrorist threats from groups such as Boko Haram, as well as monitoring agricultural production. Quite apart from the country's widespread corruption, the programme has faced difficulties linked to technological limitations: so far NASRDA has had to rely on foreign rockets to launch its satellites. China, which sees satellites as infrastructure and is investing heavily in emerging space programmes, has been keeping an eye on the country, which is Africa's largest oil producer. In 2018 the satellite operator NigComSat sold a $550 million stake to a Chinese company, China Great Wall, to finance the production of two communications satellites, and more Chinese investments are planned for the future.

Zambia

A lesser-known participant in the historical space race was Zambia's National Academy of Science, Space Research and Philosophy, founded by Edward Makuka Nkoloso in 1960. The aim was to send the first African (or 'Afronaut' to use Nkoloso's term) to the moon. The astronauts were trained at an abandoned farm not far from the capital Lusaka using homespun methods. The programme planned to launch a rocket and send seventeen-year-old Matha Mwambwa and two cats to the moon. There were also plans for a voyage to Mars in the hope that Zambians would become 'controllers of the Seventh Heaven of interstellar space'. Nkoloso said he had observed the planet through telescopes and was sure that it was inhabited by 'uncivilised' peoples. When the news of this unofficial space programme spread, journalists flocked to Zambia. All of them, or almost all, took a condescending approach mixed with a good dose of racism, treating Nkoloso like a village idiot. Only a few understood the project's symbolic significance and Nkoloso's profound criticism of the US and Soviet space programmes, which were prepared to spend unthinkable sums on a race based on pride when that money could have been better spent on Earth. Nkoloso was also a high-profile figure in the independence struggle (within UNIP, the party that governed Zambia from independence in 1964 until 1991), which he pursued with traditional revolutionary methods but also jokes at the expense of the white colonialists. And they obviously missed the joke when Nkoloso declared that the goal of the programme was to establish a Christian ministry to 'primitive' Martians.

'No spaceship has a longer record of service than the Soyuz ... Between 2011 and 2020 the steel capsule was the only means of transportation between the Earth and the International Space Station.'

talked into bringing Soviet space memorabilia under the hammer at an auction in New York in 1993. This left a sour taste in the mouths of many Russians: to these diehards their former hero had become a defector.

'NASA needs us, and we need NASA,' he defended himself.

The end of the 20th century was ruled by disillusionment. An empire had collapsed. The Russians were forced to lease their own cosmodrome from the Kazakhs. How much lower could they sink? On the run-down launch site at Baikonur in 1998, concepts like 'solidarity' and 'international cooperation' sounded just as hollow as the cylindrical modules of the ISS.

VIII

Space travel was born out of rivalry. Had people not fought each other to the death in the 20th century, no one would have been shot out of the atmosphere. Fourteen astronauts and four cosmonauts died in harness. The former were sacrificed in exploding Space Shuttles, the latter in a faulty Soyuz over Kazakhstan.

No spaceship has a longer record of service than the Soyuz. Despite two early disasters, this vessel has never been taken out of commission. Between 2011 and 2020 the steel capsule was the only means of transportation between the Earth and the ISS.

I had the pleasure of taking a seat in one once. Not in the training simulator in the forest outside Moscow but in one of the originals that had for weeks hung off the ISS like a bat. Its blackened exterior is its certificate of authenticity: during its dive back to Earth it was scorched by friction from the atmosphere before landing on the Kazakh steppe like an overheated cannonball. Because there had been a Dutch astronaut on board, André Kuipers, and the fact that relations between The Hague and Moscow had not yet soured, the steel ball was brought to the dunes of the North Sea coast as a memento. Since 2016 this bell-shaped capsule has been the pride of the European Space Agency's Space Expo museum in Noordwijk, about a half-hour's drive north of The Hague.

For my convenience, the staff have placed a small set of steps beside it. It is after closing time, and I am allowed to look around by myself. The Soyuz accommodates three adults, provided they curl themselves up like shrimps. I pull up my knees and lower myself backwards into the small central seat beneath all the buttons, dials and levers of the control panel. I shudder at the thought of what it would be like if two more passengers wriggled themselves into the capsule – and that is with the hatch still open. Everything I can see through the hatch now looks vast, even the scale model of the ISS hanging from the black ceiling of the exhibition hall. It is very cramped in here. To enter the immensity of the cosmos you first have to make yourself as small as you possibly can. The wait before take-off is two and a half hours; you lie on top of a rocket in a foetal position. The gangway has already been removed. There is no way back. There is nothing you can do but wait to be assaulted by almost unbearable G-forces.

It is after that ordeal – or so spacemen and spacewomen claim – that you reap your reward. Beyond the pull of the Earth, your torso, your head and your limbs will no longer feel the effects of gravity. If you weren't strapped in you would swiftly float out of your little seat. The concept of which way is up and which is down loses any meaning. Only those who look out of the porthole will still have the sense that the world is *below* them. For moon voyagers travelling into deeper space even this notion no longer held: they have left the Earth behind, and there is no more above and no more below.

Orbiting the Earth is child's play compared with crossing to another celestial body. In such a situation you no longer have a life raft like the Soyuz at your disposal to escape the ISS in case of emergency. The first exploration of deep space put the three passengers on Apollo 8 in a religious mood. On Christmas Eve 1968 they were the first ever to cross to the far side of the moon, where they no longer had radio contact with Houston. This complete moon orbit was when the USA started to catch up with the Soviets. To distinguish themselves from the godless cosmonauts, the astronauts showed their devotion by reading Bible texts aloud as soon as radio contact with ground control had been restored. They took turns reciting the first ten verses of Genesis:

> *In the beginning God created the heaven and the Earth. And the Earth was without form, and void; and darkness was upon the face of the deep.*

What made a permanent impression was their picture of an Earthrise (see the sidebar on page 91). The crew of Apollo 8 were the first to photograph this spectacle, in which the Earth hangs above the moonscape like half a small disc. There is no sleight of hand; the roles really are reversed. What you see makes sense yet causes your head to spin. From the perspective of the moon we live on a sphere with oceans and clouds, between which brown, yellow and green land-masses are visible. It looks precarious. Billions of souls together under a thin film of atmosphere on the spaceship that is the Earth. We hurtle through the universe with no goal or destination.

Looking at our planet from the heavens will change your perspective. At 384,400 kilometres from Earth, NASA astronaut Ed Mitchell, the sixth man on the moon, suddenly saw 'the massive insanity which had led man into deeper and deeper crises on the planet'. On the moon the necessity of 'a radical change in our culture' was impressed upon him.

To reach for the stars, I realised, exposes the human condition. The more glowing our achievements in space, the brighter the light that shines on the mess and misery on Earth. Astronauts and cosmonauts mercilessly point out the flaws of those who remain behind. Virtually all space travellers, no matter the mould in which they are cast, turn into environmental and climate activists after visiting the cosmos. Those who have been to the ISS or travelled three days further to see the moon warn that we should not make our own home uninhabitable, even if only because we have no other place to go right now.

Dutch astronaut André Kuipers, too, started attesting to the Earth's vulnerability after he was hoisted from his Soyuz

A vet, the last practising Christian in the village of Byechye in the Mezensky District, who was born at a time when organised religion was officially discouraged, decided to baptise himself a couple of years ago.

THE PASSENGER   Frank Westerman

'A stay in space tempers human recklessness. With a breakdown of muscle tissue comes a build-up of a sense of responsibility. The changes astronauts go through are generally attributed to the "overview effect".'

on 1 July 2012, weakened by half a year of zero gravity. 'We must take good care of the Earth,' he stresses time and time again. 'There is no planet B.'

A stay in space tempers human recklessness. With a breakdown of muscle tissue comes a build-up of a sense of responsibility. The changes astronauts go through are generally attributed to the 'overview effect': looking down on the Earth from space will automatically make us more inclined to pay attention to the common good than to individual concerns.

IX

Outside the Space Expo I climb the path behind the tennis courts to the Noordwijk dunes, from where you can see the North Sea. My mind is churning over a paradox. While we have not been able to put out the fires of war on Earth, we have built a peace lab in space. Is that not a form of escapism? Are we not simply trying to run away from our earthly problems by moving into space?

Behind the dunes on which I stand were the launch sites of the first long-range guided ballistic missile, the V-2. In designing this weapon of mass destruction, Wernher von Braun served a regime that had burned the books of the Jew Einstein. His knowledge and skills were to be used to help Adolf Hitler establish his Thousand Year Reich. In the underground factories where his rockets were assembled, twelve thousand forced labourers lost their lives. Magician von Braun knew everything, saw everything. It did not deter him from parading around in an SS uniform beside the leaders of the Nazi regime.

But then, after the war, his career is given an astonishing reboot. As a living war trophy, Wernher von Braun is shipped to the USA in 1945. Here he designs – this time to protect the Free World and, by his own account, Christianity – the intercontinental ballistic missile that is capable of wiping out entire Soviet cities with its atomic load. His new masters then mobilise him to fulfil the promise made by President John F. Kennedy in 1961 to put a man on the moon before the end of the decade.

Wernher von Braun does not disappoint. He gives the Americans the 110-metre-long Saturn V rocket that carries Neil Armstrong and Buzz Aldrin to the moon on 21 July 1969.

X

The Golden Records cannot be retrieved. On the NASA website you can track their distance from the sun in real time. At the time of writing, the *Voyager* that has travelled furthest is located at a distance of 22,506,674,377 kilometres; the nearer is at 18,695,510,650 kilometres. It is expected that the spacecraft will run out of power around 2025, but even after the radio signal is lost the two ships will continue to plough on through the universe. 🐦

ANDRI SNÆR MAGNASON

WE ARE
THE

A view of Jupiter, taken on 25 August 2020 by the Hubble
Space Telescope operated by NASA and ESA. The image
was captured when the planet was 653 million kilometres
from Earth. Credit: NASA, ESA, STScI, A. Simon
(Goddard Space Flight Center), M.H. Wong (University
of California, Berkeley), OPAL team

Human activity is causing effects similar to the impact of the meteorite that resulted in a mass extinction on Earth sixty-five million years ago. We're growing ever more detached from nature. While our knowledge of the universe has never been greater, we have forgotten how to appreciate a star-filled sky or listen to the silence of the night.

ASTEROID

101

W e stand on a thin crust of rock and soil on a floating ball of magma travelling around a burning sun. We are sheltered from radiation by a super-thin layer of atmosphere. Here we live and work and play and die, eventually to be recycled in this closed system, in time becoming the raw materials for something else. Soil, animals, fossils.

I yearned to go into space as a child. I longed to float in zero gravity, to see the Earth from afar, to jump around on the moon. I remember something that happened when I was twelve years old. I used to ski, and every evening I would go out to the Bláfjöll mountains near Reykjavik. Days are very short in January, so we would ski by floodlight. Once I was sitting on the ski lift surrounded by the humming noise it made and looking at people coming down the mountain. Under the lights the skiers had three shadows, which shifted as they flew between the poles on which the floodlights were mounted, creating a strange 3D effect. It was cold. I was well

ANDRI SNÆR MAGNASON is an Icelandic writer, intellectual, poet, performer, environmental activist and candidate for the 2016 Icelandic presidential elections. He has long been involved in environmental issues, a subject he discusses in *On Time and Water: A History of Our Future* (Open Letter, 2021, USA / Serpent's Tail, 2021, UK), which has been published in over thirty countries. Among his other books in English are the novel *LoveStar* (Seven Stories, 2012), the children's book *The Casket of Time* (Yonder, 2019) and *Dreamland: A Self-Help Manual for a Frightened Nation* (Citizen Press, 2008), in which he is highly critical of the rampant exploitation of Iceland's natural resources and which, in 2009, was made into a documentary film co-directed by Magnason and by Icelandic director Þorfinnur Guðnason.

SCOTT TYPALDOS is a Swiss photographer who turned to photography after completing his filmmaking studies. Since 2010 he has been researching the field of mental health, which has led him to photograph psychiatric hospitals all over the world. His work has been exhibited in numerous countries and received a number of awards.

For decades many believed all living things to be composed of stardust, and recent studies have shown that this rather romantic and poetic notion is actually fact. The basic elements of life forms – carbon, hydrogen, nitrogen, oxygen, phosphorus and sulphur – are present in the spectrum data of more than two hundred stars, albeit in different proportions from the human body. These elements are spread throughout the universe following supernovas, and the closer such events have been to the Earth, the more elements we share. Taking this as his starting point, Typaldos selected images from the NASA archive and reworked them with this concept in mind.

geared up with gloves and goggles. I was watching them rushing past when suddenly the electricity went off, causing the ski lift to shudder to an abrupt halt; at the same moment the lights went out, and the mountain was plunged into darkness. Everything went silent, and the chairs bounced softly up and down in the blackness. As we hung there in complete darkness, holding on to the bouncing chair suspended from the wire, it occurred to me that I had never noticed how loud the humming of the ski lift actually was. Our stomachs tingled. I was afraid the chair would break loose from the wire or I would be thrown into the air as we bounced upwards. For some reason, when I looked up I felt like I was looking down. This lasted just a few seconds. I had the sensation that I was looking down into the deepest abyss that I had ever encountered. I felt dizzy. There were stars between the stars between the stars that I had never before seen, and the Northern Lights blazed across the sky in a ghostly blue. In the cold stillness I understood that I was already in space. I was very small against this infinite vastness.

*

The overview effect is how astronauts describe the sense they get when they see the whole of Planet Earth from space. They see home, how the Earth is a tiny blue dot, fragile, lonely, borderless and unique. Everything becomes insignificant: politics, religion, war, everything but the deep understanding of how precious this tiny blue ball is.

The effect I felt was not the overview effect. It was a much older experience, perhaps the oldest collective experience of humankind. It was just the common experience of every one of our ancestors since before we even became human. The feeling every single child of Earth had before the era of cities and electric lighting. Now, as more and more of us are raised in cities, this feeling was no longer collective; it was a rare encounter with the largest entity of nature that surrounds us: space.

After that strange evening I always felt as if something was missing when I walked around the city, especially on dark winter days when even the Northern Lights were obscured. I felt this dome of light pollution over the city creating a barrier between us and the universe.

I belong to one of the first generations raised in the comfort of bright cities, where the stars fade as the neon lights, street lights and stadium floodlights have become ever brighter, where darkness is not part of everyday life, where you never understand even the practical importance of a full moon, where the stars are no longer a part of the world surrounding you or your experience. But I felt a bond with those who had observed the stars over the centuries. The gaze that expands the mind, leaving questions and curiosity, the desire to unlock the mysteries of what lies above us. The thought of scale and infinity that turns you into a grain of sand but at the same time fills you with the longing to understand and even the hubris to reach out and colonise the stars. I was connected to the thoughts that have driven science, maths and religion, poetry, mythology, navigation, the mapping of time and agriculture, the manufacture of instruments to help us see better.

I belong to the first generation that is disconnected from all this. We are the children of the first generation of humans to reach the moon, but we are also the first generation that is completely disconnected from space. Pollution and the lights of cities have caused the stars to fade. Most of us would be hard pressed to put names to the principal stars, and from our big cities we can hardly see them. We no longer use them to find our way around.

I learned astronomy at school, and once during term time there was a lunar eclipse. One student asked the teacher if he could tell us about it, but he said that that chapter wasn't part of the curriculum until the spring term. We might have studied astronomy, but we never went out to see the stars. I felt that that particular incident was symptomatic of our disconnected school system.

*

THE PASSENGER   Andri Snær Magnason

As a culture, we are raised with space opera, moon landings, Hubble images and knowledge that would have been unimaginable a hundred years ago, when philosophers wondered about civilisations on the moon and Mars. We are controlled and monitored from space, we receive messages every minute via satellite, get our entertainment and information, weather forecasts and GPS coordinates for our daily walks. All our movements are mapped from space to within a few centimetres. I can go on my iPhone and choose 'health', and I can see that I walked 6.3 kilometres on 6 January, 2.9 kilometres on 7 January, 2.8 kilometres on 8 January. An all-seeing eye in space knows that I was lazy on those particular days. I look to the sky and feel ashamed. Tomorrow, I promise, I will do better and walk at least ten thousand steps. Somewhere in space, circling the planet, something has been tracking me. I am not entirely sure why, but it must have something to do with money. And this money must be generated by my behaviour, my choices. The paranoia of the lunatic has come to pass. Yes, it *is* true, the satellites are tracking me, sending me messages, secretly controlling me.

It is ironic that while space has become a greater and greater part of our daily lives, when our knowledge about supernovas, black holes and the elements that make up the universe have never been richer, at the same time we, as humans, have never been so disconnected from the sky above us. My uncle was a pilot, and he would tell me about how in the 1960s they flew huge jets across the Atlantic directed by the stars, until satellites and other human-made systems took over. He came at the end of a twenty-thousand-year unbroken tradition of navigating by the stars.

*

One time I was in Greenland with some Danish architect friends of mine, travelling the 160 kilometres from Kangerlussuaq to Sisimiut on cross-country skis. I was dragging my gear behind

Composite image of the Messier 82 active galaxy. The photographs were taken by the Hubble Space Telescope, the Chandra X-Ray Observatory and the Spitzer Space Telescope.
Credit: NASA, ESA, CXC, JPL-Caltech

me on a pulka. We had split into two groups. My group was a little way ahead of the other. It was the first day, and there was a problem with my sledge, so I had to stop. It was –40 degrees Celsius, and I told the others to carry on, I would just wait for the group following behind. I waited and waited and was starting to get cold, so I decided to follow the path in the snow ahead of me. We had been given incorrect information, so the place where we had intended to stay that first night was not fifteen kilometres away as we'd been told but thirty-five. So suddenly I was walking alone, and the sun went down. But I was lucky, the moon was full, so I could see this cold landscape by its light, and I was looking at the stars and sensed again that feeling I had when the lights went off on the ski lift. I was on a planet, only ice under my feet and infinite stars above my head, and I was thinking what a shame it was that I couldn't read them. I could die because I couldn't read them or navigate by them. The weather was calm, so I was able to make out the tracks of those ahead of me clearly enough, but I knew that even a gentle breeze would cover them over in just a few minutes. I walked and called and yelled, but then I stopped. I was alone in Greenland, on the ice, walking under a full moon, and I had no gun to protect me from polar bears. I had a sleeping bag on my sledge that would keep me warm enough down to –40 degrees, so I knew that I would be able to survive at least one night. And I had some nuts and raisins in my bag. I had been walking for three hours, and I was getting a little nervous. I was doing exactly what you should never do: never leave the group, never be alone in an unknown wilderness. I continued to walk, trying to keep a particular group of stars to my left, when suddenly I looked up saw a dog sledge coming over the hill. I had always imagined a dog sledge to be a noisy affair, with barking and whipping and yelling from the musher. But it was completely silent. The dogs had soft furry feet, and the sledge glided over the hill in total silence. I was so relieved that I got the sense that what I was feeling must have been twenty thousand years old. A paleo-relief. The feeling you get when you are lost on the ice and a dog sledge silently appears under a moonlit sky.

*

We belong to one of the first generations to live without the stars above our heads; we have a fluorescent dome over our cities. The dark side of the planet is now illuminated at night. What effect does that have on our collective psyche, not being exposed to a deep, dark sky? What kind of philosophical, aesthetic, spiritual effects does that cause? Never to have looked in awe into infinity for hours at a time? What does it mean to raise the first generation of humans without this influence or stimulus? How do you measure its absence? Is it a human right to have access to the night sky? Is it comparable to being raised without music? Without colours? Without flowers?

*

I decided to do something about it. I wanted to show my children the stars, but in Iceland we have no summer stars, and going out of the city in January to look for a dark place can compromise your safety. Isn't it a human right to have access to a real deep sky?

I wrote a poem for the city of Reykjavik, a piece of 'Instruction' art inspired by Yoko Ono:

all the city lights
are turned off
from 10.00–10.30 p.m.
a renowned astronomer
talks about the stars
on national broadcast radio
everyone is invited

I first proposed this in the year 2000, and six years later the city of Reykjavik finally decided to do it, with the event taking place on 26 September 2006. It was quite stressful for me, as I knew that for that half-hour every accident, every crime, anything that happened in the city of Reykjavik during the 'Lights Out Stars On' event would be blamed on me. So the lights went out, and the strange thing was that the sound of the city went down as well. People whispered, walked around their neighbourhoods and met people in the darkness as they looked up at the sky. This hasn't been done again, but I dream of a time when major events in the sky – a big surge in the Northern Lights, a meteor shower or a lunar eclipse – will come under a protocol whereby the lights are turned off.

*

When I was reunited with those Danish architects in Greenland, they were talking about humans living on other planets. We were sitting in a tiny hunting lodge, and everyone had been assigned to read something at some point during our six-day journey. Bjarke Ingels read the opening chapter of *2312* (Orbit, 2012) by one of his favourite authors, Kim Stanley Robinson, about a future civilisation on Mercury: 'Mercury rotates so slowly that you can walk fast enough over its rocky surface to stay ahead of the dawn; and so many people do. Many have made this a way of life. They walk roughly westward, staying always ahead of the stupendous day.'

The sun boils the surface of Mercury, but there is a sweet spot that is tolerable. The city of Terminator rides on tracks that circle the planet, the heat of the sun causing the tracks to expand and thus push the city along at walking speed, keeping it in the sweet spot, completing one circumnavigation every 176 Earth days.

The architects were not focusing on Mercury, but they had been designing homes on Mars, a job they undertook for the United Arab Emirates. Mars seems to be quite suitable. The day there is twenty-four hours and forty minutes. Sounds like a dream for those addicted to the snooze button. A hundred-kilo person would weigh only around forty on Mars. They did have a few arguments while we were in the lodge, notably about whether we would be better focusing instead on living on Titan, where there is plenty of water, an atmosphere that would shelter us from radiation and hydrocarbons like methane, but Mars would be a more practical starting point. The radiation there is lethal, so you would have to live either underground or under a shelter of ice or water. So living in igloos of some kind would be one possible solution. When I see pictures sent back from Mars I do not see an alien landscape. Mostly they look like areas of the highlands of Iceland, so I do know that we would appreciate the beauty of that planet.

It is, of course, ironic that the Emirates – an empire built on the oil that is unbalancing all the Earth's systems – dream of Mars. But if you are very optimistic, then developing or even just imagining living on Mars without being dependent on fossil fuels or exploiting nature means that the same might be possible here on Earth.

But do we have to go all the way to Mars to make sensible policies for our planet?

*

Before she started school, my daughter had a nanny who was a Jehovah's Witness. The nanny gave me a pamphlet that proved there was no life anywhere else in the universe, only on Earth, because that is how God created everything. They used scientific evidence. You would need the right-sized planet (the chances are 1:1,000,000); it has to be the right distance from a sun (another one-in-a-million chance); you need a moon to stir up the elements and the right core to recycle the surface and the perfect volcanic activity, not too violent, not too slow; you need an atmosphere with the right mixture of elements that make the perfect protective shield (1:1,000,000) and the axis of 23.4 degrees that makes it perfect for seasons (1:1,000,000); and you need water (1:1,000,000); and you need a few other things with a one-in-a-million chance of occurring. So, when it all added up, the chances

An artist's impression of a quasar in a primordial galaxy (or protogalaxy), a few hundred million years after the Big Bang. Credit: ESA, Wolfram Freudling (Space Telescope-European Coordinating Facility/European Southern Observatory, Germany)

of finding another planet with all these in place was around one in infinity. It was an interesting argument, one I had not previously considered. NASA never presented me with such difficult premises. They tend to hype the existence of all kinds of planets when they need to attract more government funding.

Once my friend was getting all worked up about the chances of life on a planet in a new announcement from NASA.

I am sceptical, I said.

Why? he asked.

Well, I just read this pamphlet from the Jehovah's Witnesses ...

*

She is a nice woman, my daughter's nanny, but she did not convert me or my daughter. Many friends of mine don't answer when the Jehovah's Witnesses knock at their doors. They will only come back again and again if you are friendly and take a copy of the *Watchtower*. I made the mistake of googling the Jehovah's Witnesses. And since then the Google algorithm continued to knock on my door, offering me content relating to the Jehovah's Witnesses, trying to take me down that rabbit hole. Interesting aspect of our world. A computer trying to strengthen any belief that I may have or amplify any information I have been seeking by recommending similar information, regardless of the quality, source or truthfulness of the content. That has happened to a few people I know who dismiss the science of climate change. They were sceptical and started reading just to get ever more radical recommendations until they were lost in a web of YouTube channels and conspiracy theories and closed Facebook groups. Interesting aspect, actually. When you have a civilisation that has the power to derail a planet, you might just make an algorithm that can kill a planet. Like the most destructive virus the Earth has ever seen. A computer asteroid.

*

Elon Musk shot his Tesla into space. An odd symbolic stunt. He also dreams of colonising Mars, like Jeff Bezos and the UAE. I wrote a book in 2002 called *LoveStar*. There's a crazy entrepreneur who's something of a King Midas because everything he touches turns to gold. He creates LoveDeath, a new way to go after you die instead of the old graveyard or the oven. He shoots dead rock stars into space and makes them blaze back Earthwards as shooting stars. This turns out to be so popular that it becomes cheaper than flying with the cheapest budget airline. So when your grand-mother dies she is packed into a suit and shipped to the Love-Death facilities in the north of Iceland, from where LoveDeath rockets carry corpses into orbit. The family gathers on a hillside, and Granny comes flaming through Orion. This is deep, and it saves lots of space on Earth. About fifty million people die each year. If we shot them all into space they could form a ring around the planet about two thousand kilometres above sea level. The world produces about a hundred million new cars every year. If the average car is five metres long, then the combined cars of the world would stretch five hundred million metres, five hundred thousand kilometres, end to end. If we shot all these cars into space we could make ten circles around the planet two thousand kilometres above the planet's surface. We could make the Earth look cool, like Saturn. I wonder what we could come up with if we had a common plan to make a single object instead of all these cars. We could construct a thick steel beam to encircle the planet, just for the hell of it. The energy each car uses over a year is prob-ably enough to shoot it into space. Just try to grasp how big our industries have become. Steelmaking during medieval times, when they used swords and lances – how many tonnes of steel did they make a year? How many years would it have taken them to make a beam that could fully encircle the planet? And today, how strong are we? And is that strength our strength or our weakness?

*

To mark the International Year of Astronomy in 2009, the Hubble Space Telescope, operated by NASA and the ESA, the Spitzer Space Telescope and the Chandra X-Ray Observatory, worked together to produce an unprecedented image of the central region of our galaxy, the Milky Way. Credit: NASA/JPL-Caltech/ESA/CXC/STScI

If aliens were watching us they would think cars are one of the devices that we use to terraform the planet. They would see a dystopian world where hundreds of millions of people are forced to travel in heavy steel shells, like snails or hermit crabs on wheels, on black asphalt lanes. They would see that we need these steel shells to travel to particular destinations called workplaces, where we spend the day changing the living world into dead products. They would see that other methods of transport would be easier for us and that we do not really want those things we make because they end up piled into mountains of refuse. They would ask: What tyrant has taken control of human civilisations? Why are they using these machines to alter the atmosphere? Why are they spending all their time replacing nature with things they don't really want?

Terraforming is a term coined to describe the hypothesis that we could alter things on any planet and make it like our own Terra, Earthlike and habitable. What we are doing to our own planet might be called Marsforming, deforming, unforming or deterraforming. We are altering our planet to become less like the Earth we have known and perhaps an Earth that is not so capable of sustaining life. Around 25 per cent of all greenhouse-gas emissions come from transport. The cars we make every single year could make ten Saturn rings around our planet two thousand kilometres above the surface, and the oil they burn is like two hundred volcanic eruptions similar to the one we had in Iceland in 2010 when Eyjafjallajökull blew its top. According to estimates, the CO_2 emissions from Mount Etna are about sixteen thousand tonnes per day; the cars on the planet combined are like two thousand Mount Etnas.

*

Our planet is constantly bombarded by small meteorites, and every few million years there might be a large hit. The Chicxulub impact 65 million years ago disturbed the Earth's carbon cycle and released between six hundred and a thousand gigatonnes of CO_2 into the atmosphere, causing ocean acidification and wiping out about 75 per cent of life on Earth. Our transport fleet releases about ten gigatonnes of CO_2 every year, so over sixty years our cars will drive our Earth towards the stress levels of an asteroid impact. With all our activities combined it takes only around twenty years to put six hundred gigatonnes into the atmosphere. We are the asteroid.

*

If aliens have been checking in on us roughly every fifty years or so, their findings would be disturbing to say the least. If they came in 1820 and again in 1870 and 1930 and 1980 and 2020, they would observe a transformation beyond anything they could have imagined.

Anthropogenic mass – the total mass of all inanimate objects made by us – has now exceeded all the living biomass on Earth. Our products, our concrete, our factories and farming, and our output doubles every twenty years. Sixty per cent of all mammals on Earth are livestock, mostly pigs and cattle; 36 per cent are humans; only 4 per cent are wild animals. Seventy per cent of all birds are farmed poultry; only 30 per cent are wild. Whenever we change a habitat or colonise land for livestock farming we tend to believe that the animals simply 'go away' to other places. But that is not what happens. They just cease to exist.

What used to be a planet of diverse and abundant wildlife has become a monoculture of crops and livestock. Instead of free birds we have caged chickens surpassing the mass of all other birds; instead of lions and buffalos, elephants and zebras we have fenced-in cattle and pigs not even in sties but in cages. And our alien observers will see that we have also turned the natural laws on their heads. Apex predators are generally rare in nature. In a natural situation we have millions of roaming herbivores and a few wolves, lions and tigers that feed on their meat. Now we are nearly eight billion, and we all want to eat our lion's share, have our pound of flesh. Such a thing has never happened on this Earth before, and it is unsustainable in the long term.

What happened? the aliens might ask. And they will find out that much of the livestock farming is unnecessary in order to provide for what humans need to survive. They will write about this regime of humans on the planet, dominating all life forms and derailing the systems, not because of what's needed but because of ideas, customs, culture. They will scratch their heads over their findings. Yes, they will say, humans are intelligent to a degree, brilliant and beautiful, but their collective intelligence seems to be no different from the simplest life forms in nature. Like an algal bloom that grows to consume all its resources only to decline once it has used up those resources.

And that is kind of disappointing. Are we really that simple?

*

History happens when humans do human things. Humans make art and culture, wage wars, build empires, invent tools and religions. Mythology happens when the fundamentals are created or destroyed. The great floods, the big fires, when the sun and moon are placed in the sky. We have grown fast, and suddenly we have to realise that we have grown beyond history. What is happening now is way beyond the powers of any world leader in the past.

Mythology tells us that Moses parted the Red Sea; that is something, of course, but nothing compared with melting most of the world's glaciers and turning them into ocean. And great kings constructed fleets to navigate the great oceans, but that is nothing compared with what we are doing by altering the chemistry of the oceans, changing the pH level from 8.1 to 7.7 by the end of this century.

The leaders of the world meet to discuss things that leaders of the world have never before discussed: the temperature of the planet, how much warmer it will become. Ramses II probably believed he was descended from the sun, but I am not sure whether he believed he had the power to make the planet warmer, despite his special connection with the sun. A climate conference is more than a historic event, it is mythological; we are not talking about war or peace, religions or markets, borders and allies, we are talking instead about the future, the fundamental elements: our air, our climate, our oceans, our glaciers. Earth, wind and fire.

We are living in unprecedented times. The overview effect gave us the sense that we live on a small, fragile planet with a super-thin layer protecting us, with a delicate ocean and countries without borders. And now we have our last chance to show our intelligence, prove that we are not just mindless algae, that we can read the data, that we can imagine a future, that we can think a hundred years into that future and stop unforming our precious Terra. ✒

The clearest image ever taken of the Antennae Galaxies colliding, taken by the Hubble Space Telescope. Credit: NASA, ESA, B. Whitmore (Space Telescope Science Institute), James Long (ESA/Hubble)

The Universe Underground

What is essential is invisible to the eye ... and to all the other senses. At the Gran Sasso laboratories in central Italy, the world's largest underground research centre, scientists study the most elusive elemental particles in the universe.

PAOLO GIORDANO
Translated by Lucy Rand

A detail of the external core of the Boron Solar Neutrino Experiment, aka Borexino, an international experiment set up to study very low energy solar neutrinos at the Gran Sasso National Laboratories operated by Italy's National Institute for Nuclear Physics in Assergi near L'Aquila.

PAOLO GIORDANO is an Italian author, born in Turin in 1982. After completing his doctorate in theoretical physics, he devoted himself entirely to writing. His debut novel, *The Solitude of Prime Numbers* (Penguin, 2011, USA / Transworld, 2009, UK), which won the Strega Prize, has sold more than three million copies worldwide to date. His other titles in English are *The Human Body* (Penguin, 2014), *How Contagion Works: Science, Awareness, and Community in Times of Global Crises* (Bloomsbury, USA / Weidenfeld and Nicolson, UK, 2020) and *Heaven and Earth* (Pamela Dorman, 2020, USA / Weidenfeld and Nicolson, UK, 2021). He also writes for the *Corriere della Sera* and is one of the creators of the HBO/Sky Atlantic coming-of-age drama *We Are Who We Are*.

SAMUELE PELLECCHIA has been a photographer for more than twenty years and is the founder of the photographic agency Prospekt. He has worked all over the world as a photojournalist and in recent years has been active in visual research.

Scientists believe that dark matter comprises 90 per cent of all matter in the universe, although it has not yet been directly observed. To find out how the search for the elusive material is progressing, Samuele Pellecchia and Paolo Giordano went to the Gran Sasso National Laboratories, where the Italian National Institute of Nuclear Physics undertakes research in the field of particle physics. But discussing something that has not yet been found is no easy task, so they allowed themselves to be inspired by the stories of the researchers themselves, the atmosphere of the laboratories and the 1,400-metre mass of Dolomite rock that shields the experiments from ambient radiation.

POWERS OF TEN

It was a film that did it. A short film, to be precise, shown in the 'cinema room' at my primary school, a somewhat pretentious designation given that it was exactly like all the other rooms except the blinds were always down and there was a little cathode-ray-tube television with a VHS attached. I learned much later that the film had been made in 1977, which meant it was dated even by the time I saw it, and that it was the brainchild of the designer couple the Eameses, the ones who came up with the plastic rocking chair, the house crow and the coat rack with the little coloured balls instead of hooks. That day I noticed only the title, *Powers of Ten*, and the promise of the first screen: 'A film dealing with the relative size of things in the universe and the effect of adding another zero'.

It is early October, and a couple is enjoying a picnic on the lakeside in Chicago. The striped square of their blanket is framed from above. The man, dozy after lunch, lays down to rest, while the camera rises away from him. Every ten seconds, the off-screen voice explains, a zero will be added to the scale of what we are observing. The first frame shows an area of one metre by one metre, the second will be ten by ten, the third one hundred by one hundred and so on.

In my primary school cinema room, where you had to be absolutely silent to be able to hear, I watched as the couple in the park became increasingly small, quickly invisible, and I felt as if I had become unstuck from the ground, moving up higher and further away. Now the entire park, now all of Chicago and the great Lake Michigan, now North America and now the sphere that is the Earth, now the orbit of the moon, the solar system, and the next frame – with an aspect ratio of *ten million million metres* – includes the orbit of Pluto, at that time still considered a fully fledged planet. As the off-screen voice continues to describe what we're watching, the sun gets lost among the innumerable stars of the Milky Way until even the gaseous spiral of the Milky Way becomes just one of the countless galaxies of the cosmos. A lonely scene, the voice describes the interstellar void where lights are ever rarer, and I, as a boy, felt for a moment all the loneliness of the dark and silent space of the universe.

But it wasn't finished. Having covered the scope of the universe itself, suddenly the video plunges at dizzying speeds back to Earth, to the picnic, where the man is still asleep. From that moment on the scale of the shots closes in. One centimetre by one centimetre, one millimetre by one millimetre. The camera zooms in on the man's hand, on the irregular tessellation of his skin, the cells of which it is composed, enters the cells themselves to reveal their structure, DNA's double helix. Not satisfied, it plunges further, into the atoms, further in still, protons and neutrons, until it reaches the minimum scale conceivable by quantum mechanics. At this point I am breathless.

In the cinema room something clicks; my life is changed for ever, to the extent that later that evening I, who never tell my parents anything, would tell them all about the film. I saw that the journey into the depths of organic matter is identical to the journey out into the galaxies: a dark expanse interrupted by lights that are few and far between. As if there were entire universes contained inside every single atom of our bodies. The infinitely small looks just like the infinitely big. The microcosm like the macrocosm. I would become a physicist, and I would learn why.

RADIOPURITY

Thirty years later I am driving down the A24 from Rome towards L'Aquila on a clear January day. Stopping for fuel, I turn my face for a few moments towards the sun. It has rained continuously for a week in the city, and this light feels like a relief, the promise of something. Being a particle physicist means that when you close your eyes and enjoy the warmth of the sun, your mind can't help recalling that these rays are made of packets called photons emitted as a result of the fusion of protons inside the sun, whose pressure, together with the pressure of the gas that surrounds its core, prevents the star from collapsing in on itself. But thinking it, or *having* to think it, doesn't diminish its pleasure, and anyway you are unable to stop. It is a way of reminding yourself that you haven't lost it, even if it's been years since you worked on stars, photons and nuclear fusion.

The signpost for the section of the Gran Sasso National Laboratories that is above ground is beneath that for the Assergi CASE housing project, one of the many anti-seismic, sustainable and eco-friendly complexes built after the 2009 earthquake that devastated L'Aquila and surrounding towns, killing over three hundred people. Everything in this part of Italy preserves the memory of that event. These houses were designed to be a temporary solution, but so many people decided not to leave that they have now become permanent homes. *Home is where the heart is.*

At the laboratories I listen to the safety briefing under the raised eyebrow of gifted Italian physicist Ettore Majorana, who looks out from his familiar photograph. After this, Roberta escorts me towards the tunnel by jeep. For Roberta it was a book she read at the age of twelve, *Cosmos* by Carl Sagan. She became an astronomer,

Before becoming one of the most widely read popular science books in the world, *Cosmos* was a TV documentary, first broadcast in 1980 and exported to sixty countries, covering everything from the origins of life to humanity's place in the universe. (In 2014 the astrophysicist – and student of Sagan – Neil deGrasse Tyson presented a follow-up to the cult series, *Cosmos: A Spacetime Odyssey*.) Carl Sagan was a hugely prolific astronomer who helped shape the scientific discourse from the 1970s onwards. In his books he mainly discusses astrophysics, championing a sceptical approach to reality. He won the Pulitzer Prize in 1978 for *The Dragons of Eden* (1977) and also wrote the science-fiction novel *Contact* (1985), which inspired the 1997 film of the same name. Although he is known as an exponent of popular science, Sagan deserves credit above all for his influence on the direction taken by our space missions and the search for extraterrestrial life: he worked as a consultant for NASA and was one of the founders of the SETI (Search for Extra-Terrestrial Intelligence) project. Along with fellow astronomer Frank Drake he was responsible for the installation of the famous plaque bearing 'greetings' from the human race on the *Pioneer 10* probe as well as the Golden Records carried by the *Voyager* probes. Sagan was one of the first proponents of the theory that Titan, one of Saturn's moons, could have oceans on its surface and that Jupiter's moon Europa could have oceans of water beneath its ice crust (a hypothesis later confirmed by the *Galileo* probe), meaning that they could potentially host life.

then deviated a little and now looks after the laboratories' external relations.

Anyone who regularly travels on the motorways of Italy will almost certainly have noticed the entrance to these tunnels. The junction appears in the middle of the Gran Sasso tunnel, and after a few dozen metres leads to an imposing metal door that is always shut. Roberta tells me that the door was designed to withstand temperatures of up to 2,000 degrees Celsius for at least two hours. If a fire were to break out in the tunnel a waterfall would be activated behind the door, a 'Niagara effect' to block the fire. It is the type of door you might imagine being featured in an action film at the entrance to an establishment where research into weapons designed to wipe out humanity is taking place.

Perhaps that is why the laboratories are sometimes kept under surveillance by those who suspect that unthinkable activities are taking place within: environmental groups, no-nuke groups, anti-high-speed-rail groups; even the satirical TV show *Le Iene* and, of course, the populist Five Star political movement have been there. There was indeed an environmental accident in 2002: a spillage into the local water supply of trimethylbenzene, which is toxic to aquatic organisms. Safety protocols have since been reinforced. The damage caused was less serious than myriad other incidents that occur every day beyond the scope of public interest, but everyone assumes an air of remorse, vague guilt, if you make any reference to that event. Because this place is all about precise measurements, precision is a founding concept, and it becomes a moral attitude.

We enter the tunnel, which has walls of bare rock, dripping wet and streaked with green mould. A pumping system

Laboratories working on the Xenon1T project, set up to search
for dark matter through the use of liquid xenon.

continuously extracts the excess water. Breathable air is blown in from outside by powerful conductors that run along the motorway. The environment that Italian scientists decided to create down here in the 1980s is entirely artificial, hostile to all forms of life. I know that in this moment there is a kilometre and a half of mountain above my head, rock at my sides and, of course, below, yet I feel no sense of oppression. The tunnel is broad, and it almost immediately widens out further into high-ceilinged halls where the experiments take place, each with its own unique particle detector. For a minute I find myself imagining a huge subterranean church, a place of worship for clandestine cults.

The facilities at Gran Sasso are the largest subterranean laboratories in the world. There are others, such as Sudbury in Canada, where the scientists go down in a lift each morning alongside the miners, but they are incomparable in size. Only China has proposed building an even more impressive one, but I understand from Roberta and the rest of the staff that there's a unanimous sense of scepticism about whether, in the end, it would be a serious competitor.

Experimental physics has an innate vocation for the extreme. It creates cathedrals with impossible architecture in impossible places: satellites launched into solar orbit that form a perfect equilateral triangle at a distance of five million kilometres from one another; a 27-kilometre ring that accelerates protons under the meadows of Switzerland and France and forces them to crash into one another at dizzying speeds; sensors planted deep under the ice of the South Pole. Three years ago, in Chile, I visited the Atacama Large Millimeter/submillimeter Array (ALMA) Observatory high up in the Altiplano in the middle of the Atacama desert:

a network of radio telescopes at more than five thousand metres above sea level where the oxygen saturation is low, the ultraviolet rays penetrating and you need to cover your face and hands in sun cream regularly.

But this vocation for the extreme is not an end in itself. The physics that can be done at home or in the basements of universities, pointing telescopes out of windows or launching objects from the top of Bologna's Asinelli Tower, has been explored in its entirety. It no longer conceals surprise or discovery. Now it's about observing phenomena far away in cosmic space or hidden in the microscopic structure of atoms, frail and elusive signals, the infinitely big and the infinitely small that left me breathless while watching the Eameses' film. To succeed, purity is required. Surrounding noise must be reduced to a minimum, which, of course, includes acoustic noise but also light noise, chemical noise and any other disturbances.

Imagine having to find a tiny, faint grey dot the size of a pencil mark and in continuous motion on the wall of a building. Your only hope is that the wall is perfectly white. White, flat and uniform. That's why you have to go to the top of the mountain range to observe the signals of remote stars. That's why satellites are sent into orbit to fathom gravitational waves, hardly discernible ripples in the space–time we inhabit. And that's why these halls have been excavated, here under the most majestic mountain in the Apennines.

The purity sought at the Gran Sasso National Laboratories is of an unusual kind: it's *radiopurity*. The absence of ambient radiation. Because whenever and wherever we find ourselves, we are constantly being assaulted by radiation. Not just photons from the sun but

In TV weather forecasts 'sunshine' and 'storms' are opposites, but when it comes to space weather, the two combine to constitute a significant threat. Solar storms consist of flows of high-energy particles, the result of giant explosions on the sun's surface. When they reach Earth and come into contact with our magnetic field, they cause the aurora borealis (the Northern Lights), but the effects could also be less pleasant: disturbances to GPS signals that could put air traffic at risk or colossal internet blackouts. In our hyperconnected world the damage would be more critical than at any time in our past, and there is no shortage of catastrophic scenarios worthy of a Hollywood script or two. Unfortunately we have little in the way of historical data to study the frequency and impact of such storms, nor are we able to predict them, and solar flares, the bursts of radiation that cause them, shoot out particles, which, at close to the speed of light, travel from the sun to the Earth in just a few minutes. But ESA and NASA are working to improve their understanding of the sun, the only star we can observe at close quarters. To this end NASA's *Parker Solar Probe*, launched in 2018, will get as close as six million kilometres from the sun, seven times nearer than any other spacecraft to date, in order to study solar winds and the expulsion of coronal plasma. ESA's *Solar Orbiter*, which set out in February 2020, will take close-up photographs and gather information on unexplored solar regions, including the poles.

every other particle that comes from the cosmos and from the Earth itself follows its own fast and transparent trajectory. A constant rain of muons and pions and alpha particles and metals arriving from distant stars; the rays emitted by the traces of radioactive isotopes in the soil – thorium, uranium, radon – and those exquisitely anthropic Krypton-85 ones, an element that barely existed in nature but, following the nuclear experiments of the last century, have now filled the atmosphere. Everything emits radiation, even us. If, in the midst of all this bombardment, we want to isolate the most elusive of particles, the ones that very seldom interact – the neutrinos and particles of dark matter – then we need to eliminate all the others. We need to silence the radiation. The mountain is a good way of doing this, and this mountain in particular, with its unique rock type that forms a natural protective shell for the experiments.

Sometimes I think that this is what I miss most about particle physics: the wearying search for a purity that is not to be found anywhere else. The stubbornness with which you try to pinpoint the meaningful signals, even the most fleeting, in the cacophony of background noise. Yet at other times it seems that this demand for purity is exactly the reason I left, to throw myself into the hubbub and be absorbed by it, because maybe there, in the noise, is where real life is to be found. Anyway, I am here today. I have been wanting to come here for years, and in the artificial atmosphere of the tunnels I feel, for a moment, like I can breathe a little easier.

CATCHING THE RAYS
To study the cosmos we look through the lenses of telescopes and we send satellites and probes into space that will never return carrying messages of peace

in numerous languages to the extraterrestrial communities they might encounter. But to study the cosmos, and this is much less obvious, we must also go underground.

Nicola works on the Boron Solar Neutrino Experiment, which is known by its Italian diminutive name of Borexino. In his case it was the TV movie *I ragazzi di via Panisperna* ('The Via Panisperna Boys'), about a group of Italian scientists in the 1930s, that made the younger him giddy with excitement. But what brought him down into the laboratories was, somewhat less romantically, the earthquake. Nicola was studying physics at the University of L'Aquila, and for his degree he was working on projects that had very little to do with underground detectors, but the university building was damaged and the students were transferred temporarily to the laboratories, and he never left.

Borexino is the most successful of the experiments currently under way at Gran Sasso. In November 2020 it earned itself a place on the cover of *Nature* with the headline 'Catching the rays'. To say 'cover of *Nature*' may not sound like a big deal, but in the world of science experiments it is akin to having your portrait on the cover of *Time*. It means something along the lines of: *the most important discovery in the world out of all the discoveries in all the scientific disciplines this week is Borexino.*

The cover shows something that is now inaccessible to the eye, the inside of the detector: a metal sphere mounted with photomultipliers, lamps that make it look, in its perfect radial symmetry, a lot like the shell of a sea urchin. I admit that a part of me was hoping, against all reason, that I would not only get to see the Borexino detector but actually get to go *inside* it. A fixation I've had since I was shown a

picture of it during a cosmology course. We all have our unattainable desires.

Nicola explains, a bit too fluently, the theory that forms the basis of the experiment. He trusts that my competence is still intact after fifteen years of doing completely different things. I struggle a little because that is not the case, but I am too embarrassed to admit it, so I do my best to feign understanding as I hobble along behind his words. But at least the core of the explanation is clear to me. The sun generates light. What we call solar light is mostly light in the strictest sense of the word: gamma rays, photons emitted in the process of fusion inside the star. But fusion, which was already understood as early as the 1930s, consists of many successive steps, and there are some alternative versions. As well as photons, it produces a quantity of other energetic particles – most importantly neutrinos. To give an idea of how many: from the sun we receive around seventy billion neutrinos per square centimetre every second. *Seventy billion neutrinos per square centimetre every second.* So in the time taken to rewrite those few words some thousands of billions of neutrinos have passed through me. They have not affected me. I didn't notice them doing so because neutrinos are the least interactive of all the particles. If they do not interact, they do not manifest. If they do not manifest, it is as if, for us, they do not exist. Yet they are there. The neutrinos plough through the cosmos, through matter and through us largely undetected, indifferent.

The same cannot be said for photons. Photons interact with nuclei, electrons and with other photons, and we do indeed *feel* them. Even inside the sun they are continuously being absorbed and re-emitted and diverted. We imagine solar rays proceeding in a straight line,

THE PASSENGER   Paolo Giordano

but inside a star their journey is full of obstacles, an exhausting zigzag. One photon formed at the centre of the sun in the fusion process takes something like a million years to reach the surface. *A million years*. By the time it gets there it is already ancient light. A neutrino takes only a few seconds.

To understand what is going on inside the sun, at its core, it is neutrinos that we are most interested in. But to catch one, thirty or so a day out of the infinite billions that pass by, we have to come here, under the mountain, where all other radiation has been screened out; we have to suspend a metal sphere in a bath of distilled and extremely radiopure water; we have to work for a decade on the calibration of the sphere and its photomultipliers; we have to spend some tens of millions of euros; and when all is finally ready we have to watch every rare, tiny flicker that the neutrinos create in their passage. This is Borexino, the neutrino catcher.

HEART OF DARKNESS

If we understand little about neutrinos, we know nothing at all about dark matter, only that it exists – somewhere, maybe everywhere. We are probably immersed in it, our whole galaxy is probably immersed in it, otherwise two or three things that we do know about reality wouldn't add up. But, after years of research, dark matter is still that: *dark*. It does not manifest in any of the ways that have been suggested by science. Is it a particle? Is it heavy? Is it light? Or is it something different from a particle, something we can't even conceive? There are various experiments, here under the mountain, dedicated to hunting for this cosmic phantom. It's a strange existential condition when you reflect on it. Hours, days, years, perhaps your whole career, just waiting for a sign. Chasing ghosts.

Paolo jokes about it. He sees the humour in his days spent in the tunnel, under neon lights, where it's always the same time of day; days spent waiting for the tiniest hint of proof that dark matter exists, which he may never get because maybe it isn't where they expect it to be. For Paolo it was a person, a family friend who visited when he was fifteen and told him about what they were doing at Los Alamos. He talked of atoms. Who knows, maybe Paolo now curses that moment, although I do not dare ask him, and I can tell that physics, despite the frustrations, still gets him excited. He is swearing when I arrive because he has just 'fried the experiment' with a misstep. Talking to me, or perhaps to himself, with a hint of bitterness, he says that we can see only an imperfect copy of a perfect nature. Or vice versa; he's no longer sure.

Paolo divides his time between the CRESST and CUORE experiments. That's another trait of particle physics: finding manageable acronyms for concepts that are otherwise inaccessible. CRESST stands for *Cryogenic Rare Event Search with Super-conducting Thermometers*, CUORE for *Cryogenic Underground Observatory for Rare Events*. So they have in common the rarity of the events for which they search – dark matter in the case of CRESST, Majorana neutrinos for CUORE – and the fact of actually having a *cuore*, a heart, an extraordinarily cold heart composed of cubic or cylindrical crystals, each the size of a paperweight, crystals that are held immobile in a case. Not immobile in the (approximative) way we would normally understand it, but perfectly still to the point where the single atoms of the crystal lattice are (almost) immobile.

Can you imagine what it means to stop a *single atom*? It can only be achieved by eliminating any external interference

Above: Detail of the exterior of the laboratory of the CUORE (Cryogenic Underground Observatory for Rare Events) project, which was set up to find out whether neutrinos can be their own antiparticles.
Below: A close-up of the outer door to the laboratories.

The CUORE (Italian for 'heart') project is not the only particle physics experiment with an unusual name and certainly not the most evocative. Our most fertile hunting ground is Greek mythology: you only need to think of the DELPHI project (*Detector with Lepton, Photon and Hadron Identification*), the ZEUS particle detector, which operated on the HERA (*Hadron Elektron Ring Anlage*) particle accelerator in Germany, the ATHENA X-ray observatory, the DAFNE (*Double Annular Factory for Nice Experiments*) particle accelerator in Italy and the shameless cheat ATLAS (*A Toroidal LHC ApparatuS*) experiment. The animal kingdom offers inspiration, too, from PANDA (*AntiProton ANnihilation at DArmstadt*) experiment to CAMELS (*Cosmology and Astrophysics with MachinE Learning Simulations*). Of course, there are plenty of anonymous acronyms: CDHS, LHCF, NA60, H1 … Even when it comes to space some names are inspired, some less so: as for telescopes, we go from the mundanely descriptive LBT (*Large Binocular Telescope*), VLT (*Very Large Telescope*) and ELT (*Extremely Large Telescope*) to the certainly more evocative MAGIC (*Major Atmospheric Gamma-ray Imaging Cherenkov telescopes*). Human names are always a good choice, too, as in the SOFIA observatory (*Stratospheric Observatory For Infrared Astronomy*) and GHGSat's satellite family: Hugo, Iris and Claire. Further away, the NASA Mars 2020 project sent out to the red planet its fifth rover, eloquently named *Perseverance*, accompanied by the small helicopter *Ingenuity*. The UAE gave their first Martian probe the significant name of *Al-Amal* ('hope'), whereas China took a more poetic approach, sending *Tianwen-1* ('heavenly questions') into orbit.

and then cooling the matter to very close to absolute zero. Absolute zero, zero Kelvin, −273.15 degrees Celsius, the point at which everything becomes inert, at which the nuclei stop all intrinsic movement and electrons stop orbiting. It is the death of matter, by definition an unreachable point. We can get close, very close even. CRESST and CUORE are doing it already. They cool their crystals down to a few milli-Kelvins with injections of liquid helium and extremely complicated pumping systems.

Paolo is a master of cryogenics, an embalmer of atoms. After having momentarily 'fried' CRESST he takes me towards the heart of CUORE. As we go up the stairs, with the reverence one would have for an ancient relic he asks me to tread lightly because my clumsiness would be enough to alter the temperature of the crystals that lie metres away and behind layer upon layer of lead.

QUAKES

We tend to imagine a physics laboratory as a sterile, orderly place with white walls and screens from which lines of code rain down. It is our false belief, driven by the movies, that science is beautiful and impalpable.

In reality, physics experiments are more reminiscent of working in a garage: spanners and screwdrivers, tangles of cables, belts and karabiners, boxes full of dusty and broken testers, computer monitors that are far from state of the art and notices that say 'don't touch this' and 'don't do that'. No frivolous nod to aesthetics, equipment good enough only for its needs. The beauty is all hidden inside the detectors; in the CUORE it's in the crystals of tellurium dioxide. Beauty, in modern physics, is the intrinsic and invisible beauty of nature, which we go to extraordinary lengths to grasp, and for that beauty we do the best we can. 'Look,' Laura points, 'do you see those things that look like counterweights covered with foam and duct tape? That was my winter's work, to reduce the vibrations that we couldn't quite manage to eliminate.'

So far CUORE has not seen a single Majorana neutrino, a subspecies of neutrino the existence of which was posited by the physicist after which it is named in a visionary article that at the time was not taken entirely seriously. On the other hand, they haven't missed a single earthquake. In September 2017 I was in Querétaro, a city a couple of hours away from Mexico City. I was asleep in my hotel when an earthquake struck, and I did not wake up. I learned what had happened when I got up and spotted a huge number of messages on my phone, messages asking if I was still alive. Laura, here in L'Aquila, had been alerted by the crystals. Following the quake in Mexico – with a delay of around twenty-eight minutes, the time it took for the pressure waves to travel halfway around the planet – the temperature in the CUORE cryostat changed.

Laura is alerted every time there is a seismic quake anywhere. She is the 'Vestal of the Detector'. She is from Rome, the daughter of scientists, has a postdoc from Berkeley, and for her it was the physics teacher she had in her final year of high school. She uses words like 'isoenthalpic' effortlessly but gets embarrassed if asked to talk about herself. The first time she came to Gran Sasso for work it was winter. There was snow everywhere, and her car was a little LPG-fuelled runabout. Everything was closed after the earthquake, so she found herself stuck in a B&B halfway up the mountain. She swore to herself that she would never come back. Now she lives here. She is on call for CUORE twenty-four hours a day every day. If at three in the morning there is a quake in Croatia, she gets out of bed, connects remotely to see what's happening and, if necessary, comes down to the tunnel, alone in these gloomy ravines, to take care of 'her' detector.

I have to insist on being allowed into the cleanroom, the uncontaminated area where the pieces of CUORE are assembled. We dress in paper gowns, caps, shoe-covers. Laura fixes an oxygen sensor to her collar. The cleanroom is isolated from external air because within there are cryogenic liquids which, in contact with the air from outside, would expand so much and so quickly that it could cause anyone in the room to suffocate. Paolo, watching us from the outside, would be obliged by the safety rules not to enter.

This year the blue-zipped clothing we put on along with the shoe-coverings and latex gloves takes on a rather grim association. I stop for a moment. Laura intuits what I'm thinking and anticipates my question: 'In the spring, at the start of the lockdown, we donated all our protective equipment to hospitals in the region.' Then she shows me into the most radio-pure room to be found anywhere for hundreds and hundreds of kilometres.

FRENCH FRIES AND BUBBLE UNIVERSES

Nello gives us a ride to the surface. He comes from Turi, in Puglia, and for him it was a natural gift for mathematics. In his car he has dozens of empty plastic bottles, which form little mountains in the spaces under the seats. We cautiously rest our feet on them, but I don't dare ask what they're for.

On the journey they talk, of course, about contracts and academic positions, the most popular topic of conversation for all career scientists. But, almost distractedly, Nello drops into the conversation universes with bubbles of antimatter, leptogenesis and annihilation gases. Then, with a flash of irritation, they tell me how Thursday has always been fries day at the laboratories, something they looked forward to. The new canteen management initiated a healthy-eating campaign that nobody wanted, and when fried food disappeared off the menu, including Thursday's fries, researchers and staff staged a unified protest to get them put back on.

The impact of the light outside the tunnel is harsh. You become a slightly different animal underground, even if there for only a short time. Yet I realise that here, in the car with these guys who in another life would have been my colleagues – and despite the plastic bottles crackling under my shoes – I am starting to relax. I am at ease, I feel good. Almost as if my body recognises them, *despite everything*, as colleagues.

In my penultimate year at university my professor of field theory, the most feared of them all, came into the lecture theatre. Before starting to fill the boards with relentless and silent sums, he invited us to reflect on the fact that Gerard 't Hooft, a Dutch physicist, had won the Nobel Prize for an article published when he was twenty-five. How old were we? Already twenty-two? Was it not the right time, then, to reconsider our potential for success in this field? I always admired his frankness, I still admire it, and I do not rule out the possibility that my divergence from physics began that very morning through the discomfort that the professor induced. Yet I realise for the hundredth time, as I leave behind the steep slopes of the Gran Sasso, bright with snow, physics is not just this. It is not just successfully discovering something. If that's how I saw it then, it's because I was too young

JAMMIN'

One of the things that makes the 21st century infinitely better than the 20th is that we no longer get lost when we visit a new place. This is thanks to GPS, one of the US government's infrequent and rarely acknowledged acts of arbitrary generosity – and one of the most useful inventions to have resulted from space research. At any point the United States – whose military Space Force continues to operate the system based on a 'constellation' of thirty-one satellites – could take back its gift, paralysing many activities all around the world. As a result, Russia, China, the EU, India and Japan have developed their own alternative systems. But more than a change of heart from the USA, what worries these countries (and the Americans as well) is how easy it is to interfere with the GPS signal. During the course of various military exercises, civilian aircraft or cargo ships passing through the area have lost their GPS coordinates, and, even though they are illegal, portable GPS jammers can be found relatively easily online and are used to avoid being tracked by your boss, to fly drones in prohibited areas or even to cheat at Pokémon GO (all documented cases). The GPS signal is also used to tell the time. GPS satellites are equipped with atomic clocks calibrated to the nanosecond that are sometimes used for surprising purposes thanks to their precision: Wall Street uses GPS for timing services, particularly important in an era when milliseconds can cost you millions. States are keen to protect themselves by reducing dependence on GPS, at least for the most vulnerable infrastructure and organisations. As for the rest of us, we might just have to learn to read maps again.

The underground tunnel that gives access to the Gran Sasso National Laboratories (National Institute for Nuclear Physics) in Assergi near L'Aquila, Italy.

and too ambitious. Physics is something completely different.

With the people who are with me in this car, with Paolo, Laura, Nello, whom I didn't know until a few hours ago, I share something primigenial. A curiosity, a fascination, I don't know what to call it: the moment that is at the origin of being here. That is what I miss today. I miss that secret language. I miss the bubbles of the universe that science is made of and of which these cave laboratories are a perfect example. To live for many hours each day inside a space that is other from common existence, that is geographically and mentally other. A radiopure place, radioprotected from human complications. At least a little bit.

In 2019 I was in Oslo for the fiftieth anniversary of the first manned moon landing. At a show about the moon, in a side gallery off a corridor, I found myself in front of the short film by the Eameses on the powers of ten. It took me by surprise. I had not watched again it after that morning in the primary school cinema room, but it took just a few frames for me to recognise it. I was alone, and I watched it three times over from beginning to end until the emotion, having reached its peak, started to wane. Then I went on to the next room, and it was like stepping away from a past that I had never quite given up on.

It is often said that we should only write about what we know well, but that is not true. Because writing is about constantly giving something up, and the things that we know too well, that we respect too much, do not allow such a sacrifice. We should write only about things of which we have just a little knowledge. If I decided to come here, to talk about how we can look at the cosmos from under the ground, it is because I felt I'd forgotten enough. I felt I was sufficiently ignorant to be able, finally, to tell that story.

CATHEDRALS

One of the challenges of those who work in particle physics is never being able to explain what they do, what they study, what they're searching for, other than through unsatisfactory analogies. And it is especially difficult to explain *why* they are doing it. Why does dark matter matter? Can we extract any benefit from neutrinos? In a world that is obsessively fixated on the reasons for every action, you need a good excuse to work on the cosmos. So those who devote their lives to following gravitational waves go to great lengths to explain that without general relativity satellite navigation would not work. Those who undertake experiments on superconductivity promise that sooner or later they will improve the efficiency of microchips. And those who bash their brains over the theoretical foundations of quantum mechanics, one of the most conceptual areas of all, must explain their perversion with the promise that quantum computers will help us surf the web much faster. Partial, degrading justifications.

Even here at the laboratories at Gran Sasso they have pre-packaged justifications ready for the visitors. When the researchers feel that one of those moments has arrived, that they have run into the journalistic question of and-what-is-it-all-for, the light in their eyes is slightly dimmed. They take a deep breath and begin to lay out the official version, feigning enthusiasm. Digging holes into the mountain is necessary, yes. Developing new technology and spending tens of millions of euros on experiments on neutrinos has a practical use, of course it does. And what do we want with dark matter? Well, this is more difficult to imagine, at least for now, but at the beginning nobody knew what electromagnetic waves were for either, but hey, look at where we are now, etc. etc.

As my focus goes in and out, I wonder what if there isn't? What if there is no practical reason to be searching for neutrinos underground? If dark matter, once found, has no use? Is it not enough in itself? Does it never occur to us to ask why certain magnificent cathedrals were built, the Sagrada Familia and Notre-Dame de Paris? To understand how energy is produced inside the sun, the process that lies at the base of our existence more than any other, to understand the make-up of the cosmic halo in which we are enveloped, to describe precisely the spiral of our galaxy, to await the next supernova and be able to reconstruct the dazzling and catastrophic dynamics through the remote echo of neutrinos, is all of that not enough? Really, is it not? ✒

The final launch of the shuttle *Discovery*,
24 February 2011.

LAUREN GROFF

Night, Sleep, Death and the Stars

What drives us to want to colonise Mars? Is it right to delegate our future to the egotism of a few presumptuous billionaires? On a family trip to the Kennedy Space Center in Cape Canaveral, Florida, a place synonymous with the legend of NASA, Lauren Groff wanders through the surreal attractions of this Disney-style space theme park in search of the idealism that could inspire us to save the only home we have.

LAUREN GROFF is the author of two short story collections and four novels, including *Fates and Furies* (Riverhead, 2015, USA / William Heinemann, 2016, UK), which sold in thirty countries, featured in many best-of-year lists and was chosen by President Barack Obama as his favourite book of the year. Her most recent novel, *Matrix*, published in 2021 by Riverhead in the USA and Hutchinson Heinemann in the UK, was shortlisted for the US National Book Award for Fiction and the Andrew Carnegie Medal for Excellence in Fiction. Her work has appeared in *The New Yorker*, *The Atlantic*, *Harper's Magazine* and in numerous other publications. In 2017 Granta named her one of the best American novelists of her generation. The title of this article is taken from Walt Whitman's poem 'A Clear Midnight'.

MASSIMO SCIACCA has been working as a photographer since 1984, notably in the field of documentary photography. He won a World Press Photo award in 1998 for his work on the popular uprising following the collapse of the financial pyramid schemes in Albania. After covering the conflicts in the Balkans he worked in Indonesia, Cyprus, the Philippines, the USA, Iran, Afghanistan, Polynesia, Australia, Argentina, Chile and central Africa on reportages and documentary films.

On the occasion of the final launch of the Space Shuttle *Discovery*, Sciacca went to Cape Canaveral, Florida, to take photographs at the Kennedy Space Center Visitor Complex and some of the activities associated with what's been called 'shuttle tourism'.

From across the broad and whitecapped Indian River, the Kennedy Space Center looks like two tiny Lego sets in the distant vegetation. The palms here are windswept, the oaks are scrubby. Pelicans bob in the shallows. Eventually, one of the structures comes clear as a small and skeletal rocket launch tower, the only one visible, although we know more are hidden not too far away. The other is a square block, the Vehicle Assembly Building, where giant NASA rockets are constructed in their upright positions.

It is all disappointingly tiny; but, of course, that is only a wild trick of scale. In fact, the Vehicle Assembly Building is the tallest single-storey structure in the world at 160 metres tall, built to withstand 200 km/h hurricane-force winds. One could fit Yankee Stadium on its roof with plenty of room to spare. The four bay doors through which the shuttles are driven into the Florida sunshine are so vast that it takes them each forty-five minutes to open.

We are speeding along, but it seems to take for ever for those distant structures to come near, even when we're over the river and on to Merritt Island. In the two shining trenches dug beside the road, herons stalk and snaky anhingas spread their heavy wings to dry their feathers in the early sun.

Are we *ever* going to get there? asks my nine-year-old son. I say, Maybe!

This was meant to be a joke about our likelihood of Mars colonisation, which we had just been discussing. As usual, nobody laughs except for me.

*

I woke my family before dawn for this, scrambling them into clothing and winter jackets, warming their hands with hot bagels for a last-minute January visit to the Space Coast. The night before, it had dipped below freezing in Florida, which always feels like being assaulted by the gods. Cold in Florida feels colder than anywhere else, because one is psychologically unprepared for it. We drove through the dark prairies, still filled with the rains of hurricanes Maria and Irma from more than a hundred days earlier, and on to the Florida Turnpike ('The Less Stressway'), straight into the rising sun. The dawn was so bright and the clouds were so high and so cold that their edges were crushed in rainbows. From this natural phenomenon we get the word *iridescence,* after the Roman goddess Iris, whose powers were minor but strongly aesthetic.

Where are we going? the six-year-old said at last.

We're going to space! I said. My sons gasped with fear. I mean, I said quickly, the Kennedy Space Center.

Then the boys, who want to be either robot or rocket engineers when they grow up, glowed in the backseat. My husband is a grade-A geek who went to space camp twice when he was a kid, and he began to tell them again about SpaceX's Falcon Heavy, the most powerful operational rocket ship to date, the launch of which we have all been waiting for and waiting for, but which keeps getting pushed off. The boys already knew about the Falcon Heavy, but they listened to him anyway. It has three reusable rocket cores, the Falcon Heavy! It has twenty-seven Merlin rocket engines in it! It is seventy metres tall! First there will be a static fire test, which is a full-thrust 'wet dress rehearsal' (wet because they burn liquid propellants like oxygen!). When it launches for real, we might be able to see the launch all the way where we live in Gainesville, more than 250 kilometres away! The boys were rapt.

Space is pretty close to my secular little family's conception of heaven. It seems to open up philosophy for my husband, who calls it the Colossal Why and makes him challenge the scale of human life and accomplishment. If we had to choose a living patron saint, three out of the four of us would choose Elon Musk, the founder of SpaceX, a billionaire with the name of a fancy hybrid melon or literary villain. (Not me; I'd pray only to Anne Carson.) My boys have taken a ride in Elon Musk's Tesla Model X and were gobsmacked by the winglike doors, the trunk space in the front and the back, the power, the self-driving option. They laugh at the pun in the Boring Company, Musk's deep-drilling business. They love the idea of Musk's Hyperloop, which is proposed to use vacuum-tube technology to hurtle a train almost frictionlessly between cities. But mostly they love SpaceX, Musk's insanely ambitious private company that wants to commercialise space for the masses and some day settle a colony on Mars. I can't escape my suspicion that there's something penile about the entire concept of a space race, with its talk of rocket erections and thrust and maiden flights, and at these ages there's nothing my boy children love more than that part of their bodies.

Also, it must be said that my husband is a capitalist, and there is something in him built to adore a billionaire.

I am not; I do not. Elon Musk, to me, is a suspect personage.

I explained this in passionate monologue to which my poor husband listened patiently and children tuned out; they're used to me.

One, I said, it is ethically egregious that even a single billionaire can exist in a world in which children die of starvation every single day.

Two, do we want companies to govern space? Companies are taking over our country, and companies now have more

Christa McAuliffe was the first person to be chosen for the Teacher in Space Project that had been announced by Ronald Reagan in 1984. The idea was one of NASA's attempts to remain relevant during the difficult post-Apollo years, and it seemed to have paid off: the American public had taken to the 36-year-old mother of two, and 17 per cent of the American population watched the launch live on TV. But on that day, 28 January 1986, seventy-three seconds after take-off, the Space Shuttle *Challenger* that was supposed to take her and six other astronauts into space broke up in flight as their families looked on from the stands. It was a national trauma. The subsequent investigation determined that the immediate cause of the accident was damage to a seal, a so-called O-ring, because of the exceptionally low temperature that winter's morning. But above all, the investigatory commission cited NASA's internal culture and decision-making processes as key factors that contributed to the accident. The agency's managers had known since 1977 that the O-rings represented a potentially catastrophic defect but had not managed to address the issue adequately. They had also ignored their engineers' warnings about the dangers of going ahead with launching at temperatures below 12 degrees Celsius. After the accident the shuttle programme was suspended for two and a half years, but the culture at NASA did not change. Although the circumstances were completely different, the *Columbia* shuttle disaster in 2003 once again revealed NASA's inability to evaluate objectively the potential risks to its missions.

rights than citizens do; companies are stockpiling wealth and diminishing the workforce at the same time. The government is stealing from poor people to feed companies' greed. The entire purpose of companies is to leach the power of the government, which is to say the entire purpose of companies is to steal the power of the people. And space exploration is supposed to be such a beautiful manifestation of the people's will!

Three, the idea of colonising Mars when we are so thoroughly fucking up our own stunning planet – that gave rise to us, that supports our lives, that could, with the slightest modicum of human behavioural modification, support many more generations of life – seems even more morally fraught. Because, look! – I pointed victoriously – even in these trenches beside this road where ducks are diving and turtles have come out to bask in what pale sun there is, *look* at those bits of blown plastic, shining as though they're innocent.

To this last charge, of the stupidity of using Mars as humankind's Hail Mary when we could just divert our energy into saving the good blue Earth, my husband just shrugged and said that Musk has a plan to minimise fossil fuel use, too. His company SolarCity is the largest residential installer of solar panels. They're trying to create solar battery packs to store renewable energy.

Well. Billionaires are still immoral, I said, unable to come up with anything better.

And this is when, with his great kindness, my husband let me save face by asking my children if they'd ever want to go to Mars.

No, I said. Absolutely, yes, of course, my children said.

*

Visitors at an attraction at the Kennedy Space Center Visitor Complex.

The Space Barons

EMANUELE MENIETTI
Translated by Deborah Wassertzug

Elon Musk

Jeff Bezos

If we counted years in 'Musk time', Elon Musk would have been born the day before yesterday. In 2016 he imagined having a spaceship ready for interplanetary flights by 2018 and reaching Mars with the first astronauts five years later. It will, in fact, take much longer, and the outcomes are extremely uncertain. As a visionary – which is what we call anyone who is thinking of something else while you are speaking to them – Musk is passionate about colonising Mars. Born in 1971 in South Africa, he made his fortune in the United States in the late 1990s thanks to Silicon Valley. He was thirty-one years old when he founded SpaceX, a space company that seemed to have no future, despite the grand pronouncements of 'Musk time'. SpaceX nearly failed but rose again by becoming a NASA partner and by selling low-cost satellite launches into orbit with reusable rockets. Musk is convinced that one day trips to Mars will become routine, like taking a plane here on Earth, and that humanity must become 'multiplanetary' to ensure its survival. In the meantime he produces electric cars, solar panels and road tunnels and develops artificial intelligence systems. He believes these will improve our world and finance research into other worlds to live on. Paradoxically, it may be that what he is short of is time.

At eighteen Jeff Bezos dreamed of building 'hotels, amusement parks and colonies for two or three million people' in orbit, transforming the Earth into a gigantic nature reserve in order to protect and save it. Now that he is almost sixty and is one of the richest people on the planet, Bezos probably sees things a little differently – Earthbound consumers are crucial to Amazon's business model – but he has retained a certain fascination with space. In 2000 Bezos founded Blue Origin, a space company that would organise brief commercial space flights that would let passengers experience zero gravity, with tickets costing $200,000–300,000. In 2021 he was on the company's first crewed flight and later took actor William Shatner, Captain Kirk in the original *Star Trek*, into space. He has for some time had a passion for lunar missions: in 2013 he was pleased to announce that he had found the two engines of the Saturn V, the gigantic rocket from the Apollo programme, at the bottom of the Atlantic Ocean. Like Musk, Bezos is convinced that the future of the human race must be multiplanetary if we are to preserve Earth's resources. At the beginning of 2021 he stepped down from Amazon to spend more time on his true passion. To finance his projects he sells a percentage of shares in Amazon equivalent to a billion dollars each year. He can do this for many years to come.

Richard Branson

Had they been classmates, Bezos would have been the serious student in the first row, Musk the one distracted by what's happening outside the window and Branson would hardly ever have been in class. And yet, from a young age Branson showed what he was made of: he started his first business venture, *Student* magazine, at sixteen then turned to selling records by mail order. He founded the Virgin Group, which today controls almost four hundred companies in different sectors, from music to beverages and, naturally, space. In 2004 Branson founded Virgin Galactic to develop space tourism with a 'space plane' on which passengers experience weightlessness for a few minutes. A ticket costs $450,000, and the waiting list is already quite long. Branson was on the inaugural flight in 2021. Then there is Virgin Orbit, another company that was created to launch small satellites using a rocket launched from an aeroplane. Branson does not seem interested in making us multiplanetary; Earth is more than enough for his businesses. He finds a low Earth orbit to be an acceptable boundary.

Yuri Milner

In spite of having more than $3.5 billion in assets and investments in internet greats such as Facebook, Twitter and Spotify, we don't often hear about Yuri Milner. Born in Moscow in 1961, Milner grew up experiencing the progressive decline of the Soviet Union and its space programme, which became less and less competitive compared with that of the USA. Perhaps this is what gave him the idea to revolutionise the exploration of the solar system and the search for life elsewhere in the universe – a pursuit that had the involvement of high-profile astrophysicists, including Stephen Hawking before Hawking's death in 2018. With his initiative Breakthrough Starshot, which has cost hundreds of millions of dollars, Milner and his fellow adventurers wish to develop space probes that can travel as fast as sixty thousand kilometres per second, exploring far-off worlds that have thus far been impossible to reach in a single lifespan. They plan to use enormous sails 'pushed' by light – something theoretically possible but very difficult to put into practice. Milner is also involved in two other initiatives: one, the search for possible alien communications, and the other, a way of transmitting messages that would be understood by extraterrestrial life forms to the furthest reaches of the universe. He believes that if we were alone it would be 'such a waste of real estate'.

Into Earth orbit or higher, as at the end of 2021

■ USA   ■ USSR / Russia   ■ China   ■ Europe   ■ India   ■ Japan   ■ Other countries
— successful launches   ∿ launch failures

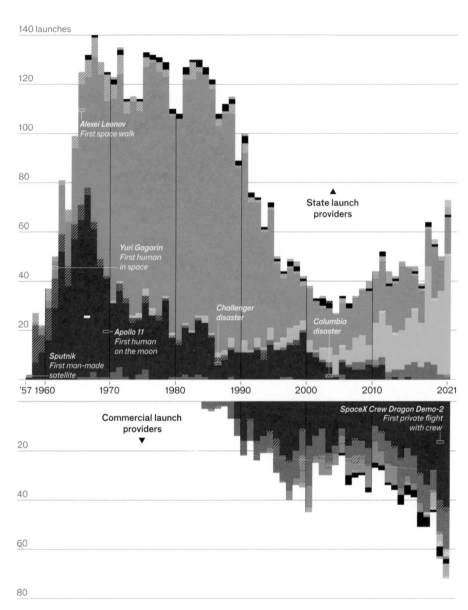

SOURCE: THE ECONOMIST, WIKIPEDIA

'To the charge of the stupidity of using Mars as humankind's Hail Mary when we could just divert our energy into saving the good blue Earth, my husband just shrugged and said that Musk has a plan to minimise fossil fuel use.'

At last we arrive. Although we are among the first tourists of the day, people are already posing and snapping photos in front of the great blue NASA globe statue at the Visitor Complex. A digital clock counts down to a pretend rocket launch that thunders the ground every few minutes.

Uh-oh, I say darkly. It's Disney: Space Adventures.

Yeah! my husband says and convinces us all to leave our coats in the car.

It is a mistake. A bomb cyclone – a vast winter storm – has covered the eastern side of the continent with ice and snow. It's 6 degrees Celsius out and the wind is blowing off the Atlantic a little way to the east, and it's Florida, so most of the lines and some of the attractions are outside. But once we've entered the park we all pretend we're too tough to complain or return to the car to fetch our ski jackets. After all, one week ago, my sons were sledging in –27-Celsius weather in New Hampshire, when a face full of powder can be dangerous.

The overweening narrative of Florida is, among other things, one of constant and oppressive heat. If even Floridians can be seduced by the story, it's because Florida is an easy state to be suckered into believing you understand. A tourist will come, see a manatee, taste some key lime pie, smell a full-grown man sweating in a Mickey suit on a 37-degree-Celsius day and go back home believing she gets the state. She probably doesn't. What she

has instead is a set of assumptions and punchlines to every other state's jokes: it's so full of retirees it's God's Waiting Room, it's so full of theme parks it's made of plastic, it's so full of people doing nutty things there's an internet meme called Florida Man, the most bumbling set of anti-superheroes in history. That day I'll notice a story on social media about a man in South Florida who sees frozen iguanas falling from trees. Where he is from, in South America, iguanas are apparently a delicacy. He will gather as many of the stiff grey beasts as he can for a feast and put them in his warm truck and drive home. As he drives, however, the iguanas will all come, horribly, back to life. Although this happened, in fact, in 2010, this man is a Florida Man.

But truth is always more ambivalent, more complicated than the simple narrative, and Florida is no exception. It's a vast state and a contradictory one, a state in which tomato pickers can live in near-slave conditions not far from a magical kingdom where small children are right now having their minds blown by mermaids and princesses. Salsa dancing and swamps, miles of fading Trump signs and world-class hospitals, alligators and NASA. If I say nobody knows Florida, I mean that there is no single Florida to know. It's all ambiguous, all so strange. And this place near Cape Canaveral seems the strangest to me, as it helped to give rise to so many of the sophisticated *things* that I love about modern humanity:

Above: Buggy and hair blowing in the wind at the Kennedy Space Center Visitor Complex, which hosts half a million visitors every year.
Below: A NASA fan before the final launch of the shuttle *Discovery* in 2011.

cochlear implants, memory foam, cell phones, CAT scans, LED lights, invisible braces, even the laptop on which I will type the final draft of this essay. Yet such fertile ground for modern invention could not possibly look any more humble, all sand dunes with spiky vegetation and teeming masses of wildlife and some scraggly looking buildings. This place holds its contradiction within itself.

*

We start exploring the Visitor Complex in the Rocket Garden where retired shuttles stand like great strange tombstones, for children to touch and run around and play in. When we quickly prove too wimpy for the chill, we go back to the beginning, indoors to the Nature and Technology building, where there are both bathrooms and dusty dioramas filled with stuffed ibises and mountain lions. Cape Canaveral is on the Atlantic coast, midway between Jacksonville and Miami, and it pushes out into the ocean like a muscular arm. Before Europeans came and thoroughly messed with the place, it was populated by the Ais, believed to have been cannibals, and the relatively unwarlike Timucua. Yet the Timucua were still warlike enough to chase off the Spanish conquistador Ponce de Léon, who first landed in Florida near where St Augustine is today. He called the landmass La Florida, or Place of Flowers. On his next stop he landed near where we are standing, and he called the place Cape Canaveral, after its thick canebrakes.

But Florida, before air-conditioning, was a stark place to live, with its heat and mosquitoes, and it was sparsely populated for the first few hundred years of European colonisation. It wasn't until 1847 that a lighthouse was built on Cape Canaveral. The lighthouse was rebuilt in 1867 as a fifty-metre iron-plated structure. A

KEEP SPACE TIDY

If one day we are able to resolve all our environmental problems there will still be a job for one of Greta Thunberg's descendants: tackling space pollution. Since the day we began to visit space we have been contaminating it with debris of varying sizes, such as broken-down satellites and probes, which can continue to orbit for decades. Space junk is a threat to future exploration and, above all, to satellites, but paradoxically the main threat to satellites comes from the rubbish generated by testing anti-satellite weapons. It is calculated that our legacy consists of twenty-nine thousand objects larger than ten centimetres (enough to destroy a satellite), 670,000 larger than a centimetre and 170 million larger than a millimetre (but still enough to cause problems for missions). Over time many of these objects disintegrate when they hit the atmosphere, but not always: some of them rain down on Earth without us even realising because they end up in the sea. The International Space Station has already had to make various manoeuvres to avoid collisions, unlike the satellites *Iridium 33* and *Kosmos-2251*, which in 2009 were involved the first frontal collision at 42,000 km/h, generating 2,200 new pieces of junk. This cascading effect, the Kessler syndrome, takes its name from the NASA consultant who warned about this risk way back in the 1970s. Since then there have been regulations and databases for 'burying' disused satellites in more distant graveyard orbits or guiding them gently into our oceans, as well as creative ideas for the spaceship breakers of the future.

naval air base was created at Banana River in the 1940s, which is why the area began hosting a rocket programme in the early 1950s. When NASA was formed in 1958, the organisation decided to stay in Brevard County, although there was little or no infrastructure there at the time – few schools, no churches, few bridges or roads – and the floods of engineers and their families had to drive over the St John's River to Orlando, then pre-Disney, a sleepy little town, to buy anything interesting. In 1961, for its manned lunar flights, NASA bought 325 square kilometres of land on Merritt Island, north and west of Cape Canaveral. Then the federal government came in and did what it did best, taming mosquitoes and transforming orange groves and building bridges and roads and putting more than $2 billion into insuring mortgages, building public facilities and schools, shoring up the airports and hospitals and creating safe harbours and waterways and water systems. Merritt Island National Wildlife Refuge was founded, conserving fifty-seven thousand hectares of land, water and swamp and preserving the habitat of endangered

If SpaceX's lofty goal is to establish a human colony on Mars, the methods it is employing to reach that goal have a distinctively materialistic feel to them – after all, it's a publicly traded company that must generate profits. The lucrative contracts with NASA and others to take astronauts and goods into orbit and beyond are a part of the equation, and so is the development of Starlink, a satellite 'constellation' that will provide commercial internet services: thousands of mass-produced small satellites in low Earth orbit that communicate with designated ground transceivers to provide rural and even remote areas with cheap broadband internet access. The reusable rockets the company has built can transport up to sixty satellites at a time (their Starship, currently under development, would raise that number to four hundred), with launches every couple of weeks. SpaceX eventually hopes to have as many as forty-two thousand of them.

To put that in perspective, at the end of 2021 there were an estimated 4,877 active satellites orbiting the Earth, already more than twice the number there had been two years earlier. When the first sixty went up in May 2019, a string of sunlit dots, many of them as bright as the brightest stars, the astronomy community gave a collective gasp. As they imagined thousands more of them (including more than three thousand launched by Starlink's main competitor, Amazon's Project Kuiper), they feared Earth-based observation of the night sky would be forever compromised. Since then SpaceX's engineers – themselves caught out by the unexpected glare – have managed to reduce their brightness, but there's not much they can do to assuage another worry, that of mounting debris. The satellites are designed both to autonomously avoid collisions and to deorbit at the end of their lives, but the sheer number of them means, according to experts, that they already represent the single greatest collision threat in low Earth orbit.

creatures, like the dusky seaside sparrow and the sea turtles who climb slowly from the sea to nest on the beaches. Into the swiftly planned and transformed area came a massive influx of educated workers: the population of the county in 1940 was 16,142, and now it is more than half a million. Despite the enormous investment, putting the space programme here was a good choice: when shuttles and rockets are launched, because of the rotation of the Earth, if anything goes wrong, it goes wrong over the ocean. Plus, the Caribbean offers plenty of places from which to monitor launches.

I will find out later that the dusky seaside sparrow has become extinct.

I am entranced by reading the history of the area until I finally notice my children sitting glumly on the fake boardwalk, eating the M&M's out of our bag of trail mix. This was not what they signed up for. Enough about the past, I say. Let's go to Mars!

*

We are alone in this strange hall, Journey to Mars, which pulsates with red and blue gels and overloud techno music. There are many shiny consoles, where all three of my males obsessively try to land pretend shuttles, and kiosks where you can take your portrait in a semi-realistic space suit and email it to yourself. In the corners, actual rovers crouch like afterthoughts: the leggy ATHLETE, the Sojourner with her solar panels, the Curiosity, the Spirit/Opportunity. All of the rovers look far more awkward in person than I'd imagined them. I am too used to technology meaning sleek design; the hand of yet another billionaire, Steve Jobs, made visible upon the world.

I learn that Mars has two moons, Phobos and Deimos, that its atmosphere is almost entirely carbon dioxide, that a year there lasts 687 days and that the planet is red because of the iron in its surface soil. Mars seems profoundly inhospitable to humans. It seems insane to even visit a place like this, let alone use it as a bank for the future of human civilisation.

Out walks a man on to a stage and – made obedient by years of schooling – we all sit to listen. His name is Ken, which is exactly right: he looks uncannily like Barbie's consort aged into a craggy sixty-year-old man with white hair. The screens behind him fill with wide-eyed evocations of a Mars landing, the mysteries and joys of colonisation. When the short film ends, Ken brings the energy. Facts shoot out of him like sparks, so many I can't possibly catch them all. Ken is a scripted proselytiser, and part of his script is to pretend he's using the screens as a touch screen to make his points. It hurts the heart a little to watch the mistimings, the gestures towards images already gone.

Ken talks of 'the search for hope'. He says that 'every dollar spent is spent on Earth'. What slowly becomes clear is that NASA is ceding a great deal of its authority to commercial interests. It has funded two programmes since it shut down its Space Shuttle programme in 2011: the Commercial Resupply Services and the Commercial Crew Program. The Commercial Resupply Services are a series of contracts with companies, including SpaceX, to resupply the International Space Station, which has been in orbit since 1998 and is now on its sixty-seventh expedition. The Commercial Crew Program was designed to prod companies into developing private space technologies for launch into low Earth orbit. (Its first flight, the SpaceX Crew-1, took off on 15 November 2020.)

Wait a second, I stage-whisper to my husband, although the music is too

'The shuttle is not as big as I'd imagined it would be ... it also looks astonishingly and erratically handcrafted, as though my quilting mother, given enough time, could make a shuttle with neater stitches.'

loud for Ken to hear me. So NASA isn't developing anything themselves any more? They're just *handing out American taxpayers' money to private profit-seeking companies?*

Shush, says my husband. He's really grooving to Ken's jam. So are the boys, who nod when Ken asks if they are going to be part of this Journey to Mars, something 'harder than anything anyone has ever done'. Something 'bigger than they are'. Talk ends. The screens go blank. Ken shuts himself down with the music. He must do this sixteen times a day.

My sons' sweet faces are filled with excitement. They quiver with it. They stand, ready to personally palpate the endless boundaries of space, and then they nearly fall over themselves to hug a robot that had come out in the middle of Ken's talk. It's a man in a glossy plastic suit gliding about on a hidden Segway, but in this place anything can happen and the boys have become true believers. Later they will be crestfallen when my husband, not really thinking, tells them it's not a real robot but rather a dressed-up man.

*

Still, we have seen the future, dusty and red and twin-mooned, and it's not even lunchtime. So we bravely return to the past. A giant orange-and-white stucco shuttle stands before the Space Shuttle *Atlantis* building, which makes it hard to

miss. Up we corkscrew to the top of the building where a movie showing us the *Atlantis* is supposed to be running, but it's glitching. So we go to see the shuttle orbiter itself, which is not as big as I'd imagined it would be, for the thirty-three missions it flew and its more than 190 million kilometres before it was retired with all the other shuttles in 2011. It also looks astonishingly and erratically hand-crafted, as though my quilting mother, given enough time, could make a shuttle with neater stitches.

The model of the control room is so analogue, no visible computers, that it makes my head hurt to imagine braving space with these joysticks and little metal toggle switches.

I see the chair, or Manned Maneuvering Unit, in which Bruce McCandless made the first untethered spacewalk, and then I see the photograph of him above the upturned blue bowl of Earth, floating far from the shuttle in the airless black of space. I have a mild manifestation of Stendhal syndrome and have to sit down, a quick plunk, on the floor.

This image, I think, is the most beauti-ful and eloquent visual metaphor for lone-liness I have ever seen.

I'm on the floor no longer than a minute before my six-year-old comes over and climbs in my lap, and I smell the musky stink of unwashed boy. I hug him and think that I never, never, never

The presenter of a US TV show during the live broadcast
of the final launch of the shuttle *Discovery*.

Night, Sleep, Death and the Stars

want either of my children to go to space, because there is more than enough loneliness here on Earth.

The nine-year-old pulls both of us up, and dizzily I follow my boys and we go down a strange slide that demonstrates – what? – a momentary lapse in gravity perhaps. And then we go into the Shuttle Launch Experience, which is supposed to give the experiencers a mild taste of what it's like to be in a launch. Up the ramp, televisions run interviews with real-life astronauts, and almost all of them are paunchy and middle aged and have either Midwestern or Southern accents.

After watching for a while, I say to my husband with wonder, Astronauts are aggressively ordinary.

Makes sense, my husband says. Their lives are a million hours of tedium and three minutes of screaming terror.

With this talk of screaming terror, the boys chicken out. Good; so, they'll never be astronauts. Some lady volunteers to take them back to the observation room where they can watch us be blown back by what we assume will be something like G-force pressures and laugh at the faces we make. I'm nervous. I hate fear.

But instead of hearing loud noises and being pancaked to our chairs, the fifty or so of us are tipped backwards and gently shaken as a rocket pulses upwards on the screen before us. Mostly I'm uncomfortably aware of the looseness in my jowls and neck. At the end, we're tipped forward as though to simulate another lapse in gravity. Above us, doors like the shuttle's payload bay open wide, and on a giant grainy screen above us we can see the spinning Earth. Someone, at some point, had thrown a soda up there. As soon as my husband points it out, I can no longer see the planet, only the brown streaks.

We're let out, and there the boys are, whole and relieved they didn't even have to pretend to be in a rocket ship for a few minutes.

The scariest thing about that, my husband says, was letting that stranger lady lead our kids away.

*

After a lunch of veggie burgers and salads, as ordinary as astronauts, we wait in the chilly wind for what we're really here for, the bus tour around the launch pads and the Vehicle Assembly Building. Every tourist in Florida has found himself here this morning. And it feels very odd that, in this day and age, many people around us are speaking something that sounds quite like Russian. I try to say this to my husband, but he is fascinated by a man wearing a red fleece, red sweatpants and white Crocs.

That's definitely an alien trying to understand earthlings, but who's not fitting in all that well, he murmurs, staring.

I'm too frozen to laugh. I grab my nine-year-old for warmth and waddle us slowly up to the bus, like an emperor penguin and its chick.

The bus is hot and full; we have to separate. Our guide is Joe, and he's a crusty bugger. He yells at us when we don't take enough pictures. I like him because halfway through the bus trip he will announce that he's a proud liberal; he'll define it to his captive audience by yelling, Liberal only means open minded! After that, the hostility will seethe off the man in the camouflage hat sitting ahead of me, and afterwards he'll give Joe a piece of his mind, but Joe will laugh it off.

We pull into the scrub; there's not much to see at first. Next to me, my six-year-old is asleep. My husband, way in the back of

There is an impenetrable mystery surrounding the disappearance of *Zuma*, a satellite launched by the US government. This is what we know: the launch of the SpaceX rocket Falcon 9, which was carrying the top-secret satellite, took place on 7 January 2018 at 8 a.m. local time from the Kennedy Space Center. The plan was for the (successful) re-entry of the first stage of Falcon 9 at Landing Zone 1, Cape Canaveral. As far as SpaceX was concerned, everything went according to plan. But *Zuma* disappeared. It is unclear what the satellite's intended purpose was or which government agency was responsible, but the clandestine nature of the mission suggests something military. We know only that it was heading for low Earth orbit and that, according to *The Washington Post*, it cost around $3.5 billion. There are few sources, and they fail to agree on details. Most believe that the satellite did not successfully achieve orbit and dropped back into the atmosphere. One theory is that it did not detach from the upper stage of Falcon 9; another that it was the second stage that caused the launch to fail. The US government added *Zuma* to the catalogue of objects in space, but the entry does not contain details of its orbit, and the official sources state that the military has nothing more to add to the satellite catalogue. Some have suggested that the satellite was actually designed to be invisible, so what better way to test it than to convince people that it isn't there? None of the many amateur astronomers who monitor satellites has ever been able to locate *Zuma*, but they found something else during their search: radio transmissions from *Image*, a NASA satellite that went missing in 2005.

the bus, falls asleep, too. But I'm entranced: space exploration takes place *here* in this motley assembly of corrugated steel buildings, concrete, parking lots, rusted metal thingies, raw wood structures, windsocks, hoses of unidentified purpose and electrified fences so that the alligators, who can climb, get zapped if they even try. I'm glad to see that the money spent is spent for the raw science of space, not for prettification.

We move around the VAB, where the scale reveals itself when Joe points out the wee little grey mouse holes, which are the three-metre-tall doors for the people to enter the building. From here, on a huge path of beige Tennessee Alabama River Rock two metres deep, the giant transporters crawl at 1.6 km/h, bringing the rockets and their umbilical structures from the VAB to the launch pads. Each 1.6 kilometre, a transporter burns 625 litres of fuel. The pads have to be so far away from the VAB and the control building, Joe explains, because no human beings can be within 5.8 kilometres of a launching rocket. The Saturn V rocket was so loud that it could easily have killed a human standing close enough. The sound can cause your organs to fail.

At last, at last, we're at the SpaceX launch pad 39A, which we can see through a chain-link fence. The service structure, a tall metal scaffolding upon a built-up beige hill, looks like the skeleton of a sad futuristic robot. The hill contains four storeys of hydraulics and control buildings and fibre-optics and power. Beyond, there's a white globe that is very cute and equally lethal, because it holds the liquid propellants. SPACEX is emblazoned on a water tower as well as on a warehouse-like building.

The tower is there, Joe explains, because just before launch it releases 1.6 million litres of water into what is called

'I carry around with me an anguish like an open wound: the knowledge that when I one day leave my little boys, I will leave them with a far worse world and darker future than the one that I was born into.'

a flame trench beneath the rocket. The water absorbs the sound and heat from the explosions. Without it, when the rockets launched, every single window in every town in Brevard County would be shattered.

As we pull away, I understand that I'm deeply, maybe bitterly, disappointed that we won't be seeing the Falcon Heavy or any SpaceX rocket at all. There was so much excitement in our house, so much discussion. But, as it turns out, we have miscalculated our dates by one day. The very next day after our visit, SpaceX launched the Falcon 9 with the secretive *Zuma* satellite in it. The *Zuma* may, or it may not, have failed; SpaceX's lips are sealed. It's all so very mysterious (see 'Whatever Happened to *Zuma*' on page 153).

*

Oh, we are tired, tired to the bones, of all the effortful imagination it takes to consider space and technology. All this *ambition,* it's so heavy. We are dumped by the busload into a theatre where we stand in the dark to watch a film. I rebel and sit on the floor with my little boy. Maybe I can take a nap. But, unexpectedly, I find myself watching with increasing attention. Among the images of the 1960s, skinny men smoking and girls gyrating in tunic dresses, comes the story of NASA.

And suddenly President Kennedy is on the screen, so young and handsome and yellowish with his bright lick of red hair, speaking of how 'space can be explored and

mastered without feeding the fires of war, without repeating the mistakes that man has made in extending his writ around this globe of ours. There is no strife, no prejudice, no national conflict in outer space as yet,' he says. 'Its hazards are hostile to us all. Its conquest deserves the best of all mankind, and its opportunity for peaceful cooperation may never come again.'

And, in the dark, surrounded by strangers, I find that I am crying. Kennedy says:

But why, some say, the moon? Why choose this as our goal? And they may well ask, why climb the highest mountain? Why, thirty-five years ago, fly the Atlantic?...

We choose to go to the moon! We choose to go to the moon in this decade and do the other things not because they are easy, but because they are hard; because that goal will serve to organize and measure the best of our energies and skills, because that challenge is one that we are willing to accept, one we are unwilling to postpone, and one we intend to win ...

I am crying because this is so beautiful, so idealistic, because this is what a real president looks like, because the year since the 2016 election has crushed my ideals under its angry boots. Kennedy's speech is responding to fear with resolute and clear-eyed courage. It's making the trip to space a challenge for humanity and not a competition.

And I'm crying, too, because perhaps,

in my fear and anger, I've utterly mistaken the foundations of space travel. It's true that humanity needs to resist something to push beyond its own complacency; this is how we've always worked, we weak and oppositional apes. And having money thrown at the starry impossibility of colonising Mars may give us, as a species, the way to perpetuate life on this ball of soil and water.

Every day I carry around with me an anguish like an open wound: the knowledge that when I one day leave my little boys, I will leave them with a far worse world and darker future than the one that I was born into. I don't know if I can forgive myself for the unconscionable act of having invited them, out of longing and loneliness, into this dying world in the first place. Kennedy suggests that maybe, instead of the capitalistic greed and preening and ego-stoking and false catharsis I've believed the Mars plans to be, they are, instead, a single blazing point of hope. Maybe they are a large-scale collaborative experiment with all realms of science and engineering, all countries of the world invited to come up with new ideas. Maybe the point is not to make a stupid terrarium on a nasty hostile planet but to show us the way to fix what we have. Maybe Mars will save Earth.

I am so moved that I can barely see the next film, projected on a screen right above a preserved real-life Saturn mission control unit. I can barely walk into the warehouse and see the Saturn rocket suspended above, so enormous, seemingly the size of a football field, all that metal to push a tiny glass pill of humans above the atmosphere. Even when we buy the astronaut ice cream, a freeze-dried vanilla sandwich, it can't bring us back to wonder. We are wondered out. We had intended to go up to see the manatees after all our immersion in space, but we're all tired, too much information sizzling the neurons into blue sparks at our brains' edges. So we pile into the car and we drive home, each thinking deeply or not thinking at all. The little boy is certainly thinking because he asks questions: How do people eat on Mars? Can a Space Shuttle do a back flip? If birds don't have mouths, how can you tell if they're happy or they're sad? Mommy, I forgot, are we driving to somewhere or away from somewhere? The sunset lights up all the grazing horses of Central Florida so their shadows stretch across the highway. We eat dinner at a delicious pizza joint in Micanopy, and the boys fall asleep in the car. The dog is ecstatic to see us. The moon is a cold cup when I walk the poor creature through the dark neighbourhood. We fall into bed quickly, and we sleep soundly and well.

*

Early in the morning the six-year-old comes down the stairs and climbs his cold little body into bed with us. We are asleep until he shivers us awake.

I had a dream, he whispers.

Mmmmm, we say, still sleeping.

It was the end of the world, he says.

We're both awake now, but we don't yet say anything. It is still dark outside, and the birds haven't yet started singing.

There was a rocket, he says. And it was big. Three hundred thousand and six hundred billion of powers. And it tried to launch, but it failed. And it blew up the world.

We all imagine this, the great final fireball. The silence goes on and on and on.

Sounds like a bad one, I can finally say. But the little guy has warmed up enough to have fallen asleep again against his father, who is also asleep, and I have said it only to myself. 🐦

All Dressed Up for Mars but Nowhere to Go

The terrain of Mauna Kea, the volcano on Hawaii's
Big Island, is similar to that found on Mars.

Josh had a dream of setting out on a one-way trip to Mars. Someone had convinced him that in a few years' time they would have developed the technology and found the funding needed to colonise the red planet. Two hundred thousand people had applied, they said. If only it had been true ...

ELMO KEEP

157

ELMO KEEP is an Australian writer. She currently lives in Sydney.

GAIA SQUARCI is an Italian photographer and filmmaker. She divides her time between Milan and New York, where she teaches at the International Center of Photography. Her work has appeared in *The New York Times*, *The New Yorker*, *Time*, *Vogue*, *The Economist* and numerous other international publications and has been exhibited all over the world.

Squarci's photographs that accompany this article were taken in the crater of the Mauna Loa volcano, Hawaii, where a NASA-funded eight-month mission called HI-SEAS (Hawaii Space Exploration Analog and Simulation) took place with the aim of monitoring the psychological responses to confinement in an otherworldly environment to prepare for future scientific journeys to Mars involving humans.

W hen Josh was ten years old he sat cross-legged on the floor in his parents' neat suburban home in Australia, enraptured. It was May 1996, and Andy Thomas had just stepped out of the Space Shuttle *Endeavour* and on to the tarmac of Runway 33 of the Kennedy Space Center. In his flight suit, bright orange against the blue of the sky, he talked in his clipped and measured, British-sounding tones about seeing his hometown of Adelaide from the godlike vantage of space. These TV images would stick in Josh's mind like gum to a boot sole.

Andy Thomas was just like Josh, Josh reckoned. He was an Australian. An Australian who'd made it all the way to NASA. Who'd been to space and back. And everyone in the world, it seemed, wanted to talk to him about it. If Andy Thomas had done it, then Josh could do it, too. That could be Josh some day, speaking before the world's media, beaming out of everyone's television, one of just over five hundred people ever to leave this planet. In that moment Josh wanted only one thing out of life: to be an astronaut.

Josh was twenty-nine when I first met him. He had been a member of the Royal Marine Commandos. An engineer. A physicist. A blast specialist, a mining technician and, briefly, a scuba instructor. He'd worked for one of the most famous artists alive. He was also a stand-up comedian – he plays Keith the Anger Management Koala, a foul-mouthed, sociopathic character in a furry suit, who provides Josh a remove from himself to exorcise a few of the demons he's been carrying around. It's a pretty weird show.

One day in 2012 Josh was sitting in an Edinburgh Starbucks, feeling down, when he came across a call for volunteers for a fledgling space programme; the application process would be open soon. There was just one catch. The mission was one way.

To Mars.

This was his shot. This was big. This was it. All these years, fostering that childhood dream. Finally, his life was going to change for ever.

When the time came, Josh filled out the form. *Could he describe a time when he had been scared? A different time when he had been stressed? Why was it important that the mission be one way?*

He paid the registration fee, uploaded a video explaining why he should be chosen for the mission and hit send.

Then he waited.

*

Mars One, a private, not-for-profit company registered in the Netherlands, might have come to your attention when it announced via press release that it had received more than two hundred thousand applications for the chance to be the first human being on the surface of Mars.

Despite not being a space-faring agency, it claimed that by 2025 it would send four colonists to the planet. Ultimately, it said, there would be at least six groups of four, a mix of men and women, who would train on Earth for ten years until they were ready to be shot into space strapped to a rocket, never to return.

It estimated the mission would cost only about $6 billion, tens if not hundreds of billions less than any manned Mars mission so far proposed by NASA. Mars One openly admitted that it was 'not an aerospace company and will not manufacture mission hardware. All equipment will be developed by third-party suppliers and integrated in established facilities.' That's how it would keep costs down, by outsourcing everything to private enterprise.

It was, essentially, a marketing campaign with two goals: first, to raise enough interest among the global community in a manned Mars mission so that crowdfunding and advertising revenues would be generated to the tune of billions of dollars; and, second, to use this money – largely to be raised through a reality television series documenting the training process and journey to Mars from Earth – to pay for the mission itself.

The mission was open to anyone in the world who wanted to volunteer. These people didn't have to have any special qualifications whatsoever; they needed only to be in robust physical and mental health and willing to undertake the mission at their own risk. As the proposed programme progressed, they would have to prove themselves adept and nimble learners, able to amass an enormous amount of new practical knowledge, not only in the high-pressure intricacies of spaceflight but also in how to perform rudimentary surgery and dentistry, how to recycle resources, how to take commands and how to maintain a harmonious team dynamic for the rest of their natural lives.

Two hundred thousand applicants would seem to suggest that the plan had solid legs – a staggering number of people willing to sacrifice their lives on Earth to take part in an open-source, crowd-powered, corporately sponsored mission into deep space. A huge amount of interest in this endeavour clearly demonstrated this right off the bat.

If only any of it were true.

*

There have been fifty unmanned missions to Mars so far. Thirty-five have failed.

Mars is freezing, –63 degrees Celsius on average, although on a hot midday at

the equator during summer it can get up to 20 degrees Celsius.

It is barren, free of features other than its frozen ice caps, vast deserts and enormous mountain peaks.

Mars is not close.

Mars has almost no atmosphere, burned off over billions of years by solar winds, leaving the surface exposed to deadly amounts of radiation. Roughly every five years the planet is blanketed in a dust storm that blocks the sun for months at a time.

For the Mars One colonists this would be home. With no way back. For ever.

*

When Mars One announced that it had received two hundred thousand applications from around the world, Josh's heart sank. That list was sure to include a bunch of fighter pilots, ex-space agency engineers, private space company employees, scientists, geologists, people with PhDs and genius IQs, even Nobel laureates – thousands of candidates far more qualified than Josh was. So when he found himself on the shortlist of people who were ready to live out their days on the lonely surface of Mars, he was shocked, to put it mildly.

Josh moved back from Britain to his parents' house in Australia to dedicate his life full time to Mars. He was going to do whatever it took to make sure he would be in the final selection. He invested everything he had – financially, emotionally, romantically, professionally – in Mars One's cause.

I had no grounds to doubt his commitment to the programme, but I also couldn't help thinking of all the things Josh had tried his hand at in his short life and found them all wanting. Going one way into space isn't something you can bail on when it turns out not to be what you want.

But maybe for Josh that was the point: having the options taken away and being left with only one purpose for the rest of your life could have been the answer he'd been looking for all these years.

*

Mars One had a core staff of only three people: Norbert Kraft, chief medical officer, Arno Wielders, chief technical officer, and Bas Lansdorp, CEO. (There were a few other employees listed on the website, but when I asked Lansdorp if those people were paid, he refused to comment.) Wielders and Lansdorp were based in their native Netherlands, while Kraft was in San Jose, where I spoke to him over Skype.

Before joining Mars One, Kraft had worked for NASA, as well as for the Russian and Japanese space agencies, where his focus was on modelling psychological testing for long-haul space flights. He was tasked with whittling down the two hundred thousand applicants – a not inconsiderable job for one person. Assessing the suitability of someone who volunteers to take a very, very slow suicide mission into space raises a dizzying array of questions. Can a person truly psychologically comprehend the hard reality of never coming back? What if the intense isolation brings on a psychotic break in one or more of the crew? How will they stave off boredom, irritation or anger in the cramped quarters of the shuttle during the several months it will take to get to Mars? And then for ever after that? What do you make of someone with a spouse and children who volunteers for the mission?

It was easy to get rid of a lot people who weren't really serious, he said, speaking in a thick Austrian accent. 'If they don't fill out their application, they're out. Or if they don't even know why they applied,

if they're asking "Is it Mars? Or is it the moon?" they're out. Anybody we didn't consider serious we considered kind of an idiot. In the videos, some applied naked. I mean, how can you come to a job interview and apply naked? So that was quite easy.'

Kraft said that some of the candidates discovered in the course of their medical checks that they were seriously ill, some with cancer, some in need of operations. 'So maybe I saved some lives there.' By May, he'd reduced the list to seven hundred or so.

'We, the people of Mars, in order to establish justice, ensure domestic tranquillity, provide for the common defence, promote the general welfare and secure the blessings of liberty to ourselves and our posterity – as well as the exploitation of our resources – do ordain and establish this Constitution for the Independent Martian Republic.' The first human crews who land on Mars to establish a colony will be representatives of the nation or the company that sent them and which, in all probability, financed their mission. But what if Mars were independent from Earth? This is the idea – common to much science-fiction writing – put forward by various political theorists and scientists, in particular Sara Bruhns and Jacob Haqq-Misra of the Blue Marble Space Institute of Science. In a 2016 article in *New Space* (vol. 4, issue 2) Haqq-Misra advocated 'liberating Mars' from Earth to allow the emergence of a 'second independent instance of human civilization'. He outlined five principles: 1) the humans who settle on Mars will relinquish their citizenship of Earth; 2) the governments and corporations of Earth cannot interfere with the politics or economics of Mars: no coercive trade, no economic interference; 3) scientific exploration may continue as long as it does not interfere with Martian civilisation; 4) the use of Martian land will be determined by the citizens of Mars; 5) everything brought from Earth to Mars will become Martian, and the Earthlings will have no rights over it. As Arnold Schwarzenegger says to himself in *Total Recall*, 'Now this is the plan: get your ass to Mars!'

All Dressed Up for Mars but Nowhere to Go

Right: Martha Lenio, commander of NASA's Hawaii Space Exploration Analog and Simulation (HI-SEAS) project, which simulates the isolation and living conditions on Mars, observes the landscape of the crater of Mauna Loa, the volcano on Hawaii's Big Island.

'Every proposal I found on the FAQ, from the landing unit to the living unit to the astronauts' suits, was theoretical. Which was somewhat putting the cart before the horse, only the cart is a pencil drawing of a toy wheelbarrow.'

I was struck by Kraft's absolute faith that this would all come off without a hitch, as though it could be made real just by believing in it.

'I want them as soon as possible to be absolutely independent from Earth. This is their goal, and they will do their own society. So you have to think it's exciting by itself to start everything from scratch: they will have their own constitution, their own laws. They have their own holidays. They definitely have different hours, but they have to really decide by themselves, and that's why they have to be such mature people to go there. You have to have the right start from the beginning.'

The details of Mars One's mission remained vague. Kraft told me that any technical questions had to be directed to Arno Wielders, who rebuffed requests for an interview, replying through the press office that he was too busy. Instead, I was directed to the website. On a page titled 'The Technology', it stated optimistically: 'No new technology developments are required to establish a human settlement on Mars. Mars One has visited major aerospace companies around the world to discuss the requirements, budget and timelines with their engineers and business developers. The current mission plan was composed on the basis of feedback received in these meetings.'

Pretty much every proposal I found on the FAQ, from the landing unit to the living unit to the astronauts' suits, was theoretical. Which was somewhat putting the cart before the horse, only the cart is a pencil drawing of a toy wheelbarrow.

*

No human being has left low Earth orbit since the last Apollo mission in 1972, and the effects of long-term space travel is not a vast topic of scientific medical literature.

The longest any person has spent in space was the fourteen months cosmonaut Valeri Polyakov lived on the now-decommissioned *Mir* space station; another cosmonaut, Valentin Lebedev, spent 211 days in orbit in 1982, during which the elevated radiation levels resulted in his losing his eyesight to cataracts. The flight to Mars is projected to take between seven and nine months.

Exposure to galactic cosmic rays increases the likelihood of cancer and Alzheimer's as well as suppressing the immune system. Building a craft capable of insulating astronauts from such deep-space radiation (including lethal amounts from solar flares that can erupt without warning) while finding a way to keep the craft light enough to be able to carry sufficient fuel, remains a work in progress.

Zero gravity has a deleterious effect on the human body; over the course of a trip to Mars, it could result in a loss of 20 per cent of muscle mass total and the loss of 1.5 per cent bone density per month. To mitigate these effects, astronauts on long-haul missions usually

engage in a rigorous, tethered exercise regime.

Surface gravity on Mars is only 38 per cent that of Earth. What this would mean for the long-term health of colonists on Mars is not known.

How the colonists might cope with a deficiency in vitamin D from a lack of sunlight, however, is. Vitamin D deficiency can also cause loss of muscle and bone density, can suppress immune strength and, at its most severe, causes blindness. The same goes for the intracranial pressure zero gravity places on the human eyeball.

Sleep patterns are badly disturbed by space travel, and more than half of astronauts on long-haul missions take sedatives to help them sleep. Fatigue and lethargy result in impaired cognitive function and an increase in critical errors, which is why astronauts only have six and a half 'fit' work hours per day.

A lack of energy can be exacerbated by the limited diet on which astronauts must subsist. Once their initial supplies ran out, Mars colonists would eat only food they could grow themselves, a plant-based diet, augmented by legumes and maybe insects.

Depression, anxiety, listlessness, hallucinations and chronic stress have all been reported in live missions and training simulations – as have communication breakdowns and conflict among crew members and between crew and mission command.

A well-known effect on astronauts on long missions is the dip at the halfway point, when the excitement has worn off and the return home seems unbearably distant. There is no way to know how the human mind will encounter passing the threshold of no return, when the Earth recedes from sight and the pitch-black enormity of deep space and the impossibility of ever turning back sinks in.

Eventually the four Mars One colonists will arrive on an inhospitable alien world, with only themselves for company for two years, until another flight with four colonists is planned for arrival – if they, too, survive the perilous trip through the vacuum of space. They will never speak to anyone but one another in real time ever again; the delay in relaying communications between Mars and Earth is twenty minutes, minimum.

They would be the most isolated human beings in history – a mantle currently held by Michael Collins, who orbited the dark side of the moon, alone, in 1969, although he has said of his incredible solo journey that he never felt lonely.

*

David Willson is one of the few people inside NASA who cautiously thought what Mars One was doing was 'kind of cool'. He's Australian, too, and an unabashed nerd who proudly points his webcam around the walls of his office to show me his framed *Star Trek* posters and faked UFO photographs on his mantle.

'It's a chicken-or-egg proposition. What Mars One is trying to do is to be the egg that attracts chickens,' he told me. 'If they create a market for the development of this technology, then private companies will be racing to get the costs down to fulfil the demands of the market.'

But the technical challenges of getting people out of low Earth orbit and into deep space remain vexing. Then there's the problem of what happens after they arrive on Mars.

'They're going to be living like moles,' Willson said. 'I don't think that the people who volunteered really appreciate that they're going to spend the rest of their lives living in a submarine.'

The first colonists would probably spend most of their time repairing the equipment that is keeping them alive. 'Replacing parts and replacing a toothbrush, having toilet paper – there are some things that modern society expects and does, and there would be significant degrading of your lifestyle on Mars,' said Willson. 'You would probably end up living like we did in the 18th century. With much simpler equipment, much simpler kitchen tools, much simpler things in all respects. It might be a lot like going back in time. Storing food is very important to survival, obviously, and people used big containers before and filled them up with vinegar, like with eggs. You would have been able to store half a year's food in these primitive ways without a freezer. You could have a freezer on Mars, that's not the problem; it's that you would have to be able to repair it, and if you can't, you're dead.

'Just the landing on Mars, if it hasn't been done before, is going to be a big, big, big thing. But if you're going there for ever to establish a base, that's just one tiny step on the longer journey. And each step is going to be breaking new ground. Every day. First to bury a base. First to grow food on Mars. Do we really know that food can be grown on Mars using its available resources? We don't.

'Another hurdle is dust, which is quite fine on Mars, and it would not be good if that got into your lungs. There are also chlorates, about half a per cent in the dust, and chlorates shut down your thyroid gland. Then there's the radiation environment, which if they stay underground can be dealt with.

'We don't know what else will happen. We don't know what your medical condition will be after five years. You might not be *able* to return to Earth after five years. You might have to undergo body acclimatisation in a rotating space station for several years before you could come back to Earth. We don't know. It's a question. We don't know, we don't know, we don't know.'

*

Imagine never feeling fresh air on your face again, never being warmed by the sun's life-giving rays, never hearing an orchestra play, never feeling whisky sink warm into your chest. No more walking on grass in bare feet, inhaling the scent of the air after a storm, watching kids play elaborate games in their secret worlds. No leisure time. No loved ones. No hope of release. No freedom to roam. No variation in a practically tasteless diet. No sex with the person you deeply love. Cramped quarters. Limited showers. An unrelenting work schedule. Darkness. Isolation. An ineffective sleep schedule. Constant fear, chronic stress and hypervigilance. The ever-present threat of death.

*

Eventually I was able to speak with the public face of Mars One, co-founder and CEO Bas Lansdorp. We connected as he was waiting to catch a plane back to the Netherlands. 'I'll try to keep my voice down a bit,' he said.

Lansdorp's background is in wind energy, but he billed his professional areas of expertise as 'entrepreneurship, public speaking, startups'. I asked why for him Mars One's mission was something so important to achieve, even though he said many times that he personally would not want to ever go to the planet himself.

'Mars One sounded like some kind of intergalactic Uber. If checks and balances are too expensive, just do away with them in private enterprise.'

'For the world at the moment a mission to Mars is exactly what we need. I think it can give us a common goal, something to aspire to together, something to work on together, something to unite us. Getting young kids excited about space exploration, having astronauts as heroes instead of pop stars. But I think the bigger picture of having a common goal, a dot on the horizon, that's the most important thing.'

Global unity is the goal? I asked.

'To be honest, for me it's not an important part of the programme. For me it's just about the goal of getting this done,' he said, confusingly. 'For me personally it's really about building the next base for humans to go to. For me it's more the technology challenge and the building challenge than the actual big picture.'

I asked how he saw Mars One being able to get a mission to launch for such a comparatively small budget compared with NASA's proposals for a manned return mission.

'That's a question I get a lot, as you can imagine. There's a few different factors. First, Mars One is a private organisation. We have no political obligations, which means we can just find the best supplier for the best price. In NASA they have a problem where if they do a mission like the *Curiosity* rover, each component has to come from different states because each state contributes to NASA, and they want their money back, basically. Mars One doesn't have that problem.'

(NASA's David Willson said, 'No, I don't think this is correct. Choosing vendors is a very serious business, and NASA – and the US government – does not want to be seen favouring anyone. Competitive selection is based on many factors as well as cost.')

'Another reason is that the space agencies have become too risk-averse, which is extremely expensive. It's not at all allowed for anything to go wrong, which takes a lot of paperwork to ensure. While we think having a little bit higher risk with the mission is very acceptable and is something that will reduce costs significantly.'

This made Mars One sound like some kind of intergalactic Uber. If checks and balances are too expensive, just do away with them in private enterprise. If people might die in the course of your mission, just have them sign a waiver. Neither NASA nor the European Space Agency nor the Japanese Aerospace Exploration Agency nor any government body can knowingly allow its citizens to die in space, and they certainly can't send them one way with no way to retrieve them.

Mars One listed SpaceX on its suppliers' page, but SpaceX had no current contracts with Mars One and said as much via email, adding that the company was always open to future contracts from all interested parties. The contracts Mars One did have were with Lockheed Martin, which was undertaking a feasibility study for an unmanned craft based on their 2008 unmanned *Phoenix* lander mission,

but which couldn't be completed until it received the payload specifications from Mars One – specifications that Mars One had only recently put out a call to universities to provide. Lockheed Martin confirmed that this contract was under way and that it was waiting to receive the payload specs.

Another contract Mars One had was with a company owned by Endemol, which produced *Big Brother*. The two companies announced the deal in a joint press release, but when asked for comment Endemol wouldn't confirm whether or not the production was for a pilot or for a fully commissioned series, their PR director writing: 'Things are at a very early stage, and we're not yet in a position to add anything further to what was detailed in the press release.'

A reality television series was the lynchpin of Mars One's plan. It was through this that it intended to raise the necessary capital to actually fund the mission via advertising revenue and broadcast rights. The proposal was to film the final candidates 24/7 for the duration of their ten-year training mission on Earth, from the selection process to lift-off, and then to continue to broadcast the mission itself, live, beamed in perpetuity back from Mars to viewers on Earth.

When we spoke, there was no network buyer for the show.

For the rights to advertise and screen this *Survivor* in space, Mars One estimated revenues of upwards of $8 billion, basing its estimates on the revenues of the London 2012 Olympic Games as well as projections for Rio de Janeiro 2016. With this money, Mars One would then be able to purchase the space-faring technologies that, ten years later, companies like SpaceX would have perfected, ready to send the Mars One astronauts on their journey. There was the small problem of not having the money until you had the show and not having the show until you had the technology and not having the technology until you had the money and possibly not having the technology in time, or ever, which was why, Mars One said on its website, the schedule was flexible.

*

CEO Bas Lansdorp was vague on the details of the company's financial status. On its site, Mars One listed a tally of where its donations were up to, split by country, all the way down to Bosnia and Herzegovina's $1 pledge. The total was at the time $633,440 (about half of which may have come from an IndieGoGo campaign, which was intended to raise $400,000 but with just over eight thousand backers stalled at $313,744). That was just over 0.01 per cent of the $6 billion mission price tag.

'Right now, Mars One is receiving funding from investors, from donations from all over the world and from small corporate sponsors that are helping us. We're in negotiation with a few very large brands, if they're interested in partnering with Mars One,' said Lansdorp. But he wouldn't talk specifics about which brands or what their investment would be. He said the company currently had the funds

Colonising Mars is a terrifying challenge in so many different ways. Even lay people who don't fully understand the technical difficulties can easily imagine the psychological aspects: you need only to have worked in an office for any length of time or taken a very long car journey to get it. To study the human dynamics in small groups over long periods, anthropologists have turned to accounts of sea voyages, from Christopher Columbus to the polar expeditions of the 19th and 20th centuries, to try to understand what influenced their success or failure. The forced isolation in the research bases of the Antarctic or in submarines has provided useful clues, but specially designed experiments have also been undertaken. The first Martian habitat simulators were created by the Mars Society, a nonprofit organisation, but Russia was, for many years, the front-runner in such studies, perhaps because there was a legacy of missions being aborted following crew issues. The longest simulation ever was carried out at a Moscow laboratory: 520 days, six volunteers, all men (in at least one previous experiment a woman had been subjected to sexual advances, and the researchers must have concluded that the problem was the woman). Between 2013 and 2018 NASA carried out six missions at the HI-SEAS facility on a Hawaiian volcano, which were suspended when, on the fourth day of the sixth simulation, a short-circuit electrocuted one of the volunteers and the others started arguing about whether to call for help. The photographs illustrating this article are taken from one of the HI-SEAS missions, and you can also listen to a fascinating 2018 podcast entitled *The Habitat*.

available for the unmanned lander feasibility study and the suit study, which were both under way.

Then there was the company's claim that two hundred thousand people had applied for a one-way ticket. This incredible piece of information issued by Mars One's press office was picked up with credulous haste by news outlets around the world. Even religious leaders made their opinions known, with the UAE-based General Authority of Islamic Affairs and Endowment strictly forbidding Muslims from applying, as to leave the sanctity of Earth was an affront to Allah.

But Norbert Kraft, the chief medical officer, told the *Guardian* he was sorting through eighty thousand applicants, not two hundred thousand. NBC News

Living in Space

EMANUELE MENIETTI
Translated by Deborah Wassertzug

The body

Yuri Gagarin, who in 1961 became the first human in space, did not find his mission that difficult, saying, 'The effect of weightlessness has no influence on the normal state of the organism and on the physiological functions of a human organism.' Sixty years on we now know this not to be true. The most comprehensive experiment on the effects of the space environment on the human body was conducted with the involvement of Scott and Mark Kelly, identical twin brothers who were both NASA astronauts. From 2015 to 2016 Mark stayed on Earth while Scott spent a year in orbit aboard the International Space Station. In comparing them, researchers noticed differences, most of which were temporary. Weightlessness makes it easier to move, so muscles lose tone, strength and volume. Bones become less dense and more fragile. This can be remedied through exercise using equipment that encourages physical exertion in an environment where heavy objects can be moved with the mere touch of a finger. Weightlessness has an effect on vision as well: eyeballs are flattened slightly and become swollen, causing vision problems that resolve in time once we return to Earth. Furthermore, blood flow to the skull increases, which often causes astronauts faces to look bluish and somewhat swollen during their first days in orbit.

The mind

In 1973 the crew of the *Skylab* space station conducted a one-day strike to protest at the excessive work schedule imposed by NASA. Later the astronauts involved joked about it, and to this day the details of that presumed mutiny in space are not clear. But a crew that can rebel against instructions sent from Earth is something that space agencies are well aware of. For this reason, future astronauts go through many screenings, tests and training regimes together, which address the psychological aspects of life in space. Crew members undertake training together to get to know each other and strengthen their bonds. The idea isn't necessarily that they become best friends, but rather the goal is for them to know they can count on their colleagues. This system works, and long orbital missions have not had any major problems on that score, save for the kinds of minor disagreements that might occur in any work environment. Days are long and filled with duties and responsibilities that reduce the risk of apathy caused by a reduction in social interactions and by life lived in the tight confines of a space station. The ability to communicate with friends and family back on Earth reduces the emotional burden, helping to create one's own space in space.

Life on board

Living on the International Space Station for months is an incredible experience, but without adequate preparation it is very stressful. Just think how you feel after a long-haul flight. Now think of living for months shut up inside an aeroplane with very few other people, having to perform maintenance and work on board, while a day goes by from dawn to dusk in just ninety minutes. There is no difference between up and down. The modules where the astronauts live and work are covered with lights, hooks, Velcro and computers that produce a constant hum. They sleep in sleeping bags that are anchored to the wall so they don't float around the capsule, inside small berths that look like old telephone boxes. There are moistened towelettes within reach to keep surfaces clean and prevent water from floating around, but also for astronauts to wash themselves. They mostly eat dehydrated meals. Their bathroom system recycles urine into potable water and looks more like a vacuum cleaner than a toilet to avoid unpleasant incidents. During the Apollo 10 mission the crew noted that something had escaped following one astronaut's visit to the bathroom. 'There's a turd floating through the air,' said mission commander Thomas Stafford in a matter-of-fact voice.

Sexuality

We have never heard of sexual relations between humans in space, but there is strong scientific interest in the subject, given that couples might enlist for a hypothetical long trip to colonise Mars. Having sex in zero gravity creates some complications for movement. Some have proposed using a sort of cocoon made for two, with handles to keep partners from floating away at a critical moment. In 1973 author Isaac Asimov proposed space sex as an alternative to exercise for astronauts: 'It would be a necessary therapy, but – naturally – there would be no need to even mention it.' On NASA's vast website, the word 'masturbation' appears in only one document from 1971. The agency has never dealt with the issue publicly, but several astronauts have made it clear that they had their private moments while in orbit. One Russian cosmonaut spoke of having adult films sent from Earth for inspiration. In recent years space agencies have worked to reduce gender discrimination on Earth and in orbit, but women astronauts must still wear suits and equipment designed for men, making it difficult to deal with specifically female needs, including menstruation.

'Why do human beings need to colonise the surface of Mars? What could possibly drive someone to leave the Earth behind for ever to die on a barren rock in the frozen depths of space?'

tallied the number of video applicants on the Mars One website and came to 2,782, each of whom paid an application fee of between $5 and $75. I asked Lansdorp if in the course of fact-checking this story he would allow me to see the list to verify the number. I asked where the two hundred thousand people registered their interest and if it was ever made public. His answer was ... complicated.

'I don't know if that was ever made public, but they have registered on our website for applying for our programme,' he said. 'Then there were a number of steps where people had the opportunity to drop out as that was exactly the point. The application process was kind of a self-selection that avoided us having to review all of them. The first step was paying the application fee. A number of people already dropped out there. Then there was a video that you had to make and questions that you had to answer. And that's also where a lot of people dropped out, that they're not lying in their motivation.'

I asked again if it would be possible to share the list in order to verify the figure.

'Of course we cannot share the details of the applicants with you because that's confidential, private information that we cannot share.'

I suggested that the names could be redacted to maintain the privacy of the applicants before viewing the list.

'Ah, no. I'm not interested in sharing that information with you.'

He emailed later, with an invitation to come at my own expense to Mars One's office in the Netherlands and see the list in person, although cameras would not be allowed. 'I will need to read your article before publication and reserve the right to deny you access to the list if I don't like what you wrote.'

I told him that, of course, that wouldn't be possible.

I worked on this story for several years, and when I began I wanted to understand the most obvious question of all: why? Why do human beings need to colonise the surface of Mars? What could possibly drive someone to leave the Earth behind for ever to die on a barren rock in the frozen depths of space? What tangible good would come to us in spending tens – even hundreds – of billions of dollars on sending a tiny group of people to live 18th-century lives there? Could we not find more effective ways to spend that money here if the ultimate goal is protecting the future of the human race?

I was told by Bas Lansdorp that society has lost its way and that the young people of the world should idolise explorers, not pop stars. I listened to David Willson explain that if we were able to prove that life exists on Mars, however primitively, we would have to profoundly reassess what it meant to think we were the only

2021 was, arguably, year zero for space tourism. After repeated claims and false starts, Richard Branson's Virgin Galactic and Jeff Bezos's Blue Origin finally took civilians into space – although the definition of space is open to interpretation. Prior to this, only seven private citizens, paying their own way, had gone into space. The pioneer was Dennis Tito, an American, who in 2001 reportedly gave $20 million to Space Adventures for transport on a Soyuz and accommodation at the ISS. But that line of business dried up when NASA abandoned the Space Shuttle programme and block-bought all the available slots on Soyuz, so for many years there were no seats to be had for leisure travellers. Axiom picked up where Space Adventures left off, sending its first tourists to the ISS on a SpaceX rocket in 2022, but the price, $35 million, remains exorbitant. What Virgin and Blue Origin do promise, however, is to bring space tourism – in the form of suborbital flights, including a short period of weightlessness – if not to the masses, at least to a greater number of the elite: tickets for the Virgin Space Ship *Unity* are a (relative) snip at $450,000. Prices should go down for proper orbital trips once SpaceX's Starship and Boeing's much-delayed Starliner are ready: patience has always been the most important virtue for wannabe space tourists. Meanwhile, the hospitality infrastructure is being built. Axiom is developing modules – including a Philippe Starck-designed glass-walled panoramic cupola that will be attached temporarily to the ISS and eventually reconfigured into a brand-new space station – and Bezos is planning his own Orbital Reef. Many more projects have been floated and then scrapped, but momentum is gathering, and more regular people – not just billionaires – have now 'been there, done that'.

living things in the universe. I read Elon Musk's half-joking comment 'Fuck Earth! Who cares about Earth?' I interviewed people at government space agencies who are geniuses. I came away from this in complete awe of the NASA people's minds (not so the minds behind Mars One), and with an appreciation of why space exploration is so very expensive: because it is incredibly difficult and incredibly dangerous.

Here is what I also think: what really drives this enterprise is the ancient, unbearable anxiety of death. Building a colonial outpost on Mars is a quest for immortality – to live on in human history for those who succeed, to stave off the inevitable death of our species, at least for those who believe we can upload human consciousness to a frequency sent out into space, or just recreate our unlikely habitat

THE PASSENGER   Elmo Keep

Above: Space suits photographed inside the dome on Mauna Loa, where six researchers lived in isolation for eight months. During that period the researchers could leave the dome only for brief excursions, wearing space suits with helmets and using an oxygen supply.
Left: Researcher Jocelyn Dunn looks out of the dome in which she lived with five colleagues in the crater of Mauna Loa, Big Island, Hawaii.

on another planet which happens to be next to us.

*

If you are ever standing in the very narrow path of the shadow of a total solar eclipse, when the moon passes between the Earth and the sun in a configuration known as syzygy, you will see the stunning effect of the corona of the sun flaring from behind the moon in broad daylight, as the circumference of the moon fits precisely inside the circumference of the sun from our terrestrial perspective.

This happens only because the sun's distance from Earth is roughly four hundred times the moon's distance, and the sun's diameter is likewise four hundred times the diameter of the moon: an almost exact ratio.

The odds of this configuration occurring anywhere in the universe, least of all at a place and time in which intelligent, self-aware life is present to observe it, are so minuscule as to be incalculable.

Six hundred million years from now, Earth's tides will have pushed the moon too far away from the planet for total solar eclipses to be possible any longer.

If all human life were to disappear from the Earth tomorrow, it would take the planet only a hundred million years to completely reclaim the surface, leaving no single trace of proof that intelligent beings ever existed here. All the satellites orbiting the planet would fall, untended, many coming to rest at the bottom of the sea.

The last human-made structures standing would be the Pyramids and Mount Rushmore; its granite resists erosion at an elevation that exposes it to little wind, leaving it recognisable ten thousand years from now. In five million years it will be gone.

In 7.6 million years' time, Mars' moon Phobos will have come close enough to the planet's surface to be destroyed by gravity and torn into a ring that will then orbit the planet for three million years, after which the debris will smash into the face of our best hope for repopulating the solar system.

In five billion years from now our sun will enter its red-giant phase and expand to at least two hundred times its current size, enveloping Mercury, Venus and quite possibly the Earth in the process.

In one hundred trillion years all the hydrogen in the universe will be exhausted, and so all remaining stars will die. In one hundred vigintillion years quantum tunnelling will turn all matter left in the universe into liquid.

In $10^{10^{120}}$ years (zeros are now added in septillions, numbers too big for our minds to grasp) our universe will experience its heat death, encountering maximum entropy when there is no longer enough thermodynamic free energy to sustain processes that consume energy – like life.

By this point, time itself will have ceased to exist.

You can right now, if you like, float gently and lovingly over the Earth and take in the view from the International Space Station (visit https://video.ibm.com/channel/live-iss-stream). Any of us with an internet connection can get a low-fi insight into what astronauts call the overview effect, the feeling of seeing the majesty of Earth from space and trying to take in the enormity of it and the tiny unlikeliness of yourself.

You may find it pleasantly reassuring.

*

I knew that I would have to tell Josh about all this.

'If all human life were to disappear from the Earth tomorrow, it would take the planet only a hundred million years to completely reclaim the surface, leaving no single trace of proof that intelligent beings ever existed here.'

I would have to tell him that from everything I could find, Mars One didn't appear to be in any way qualified to carry off the biggest, most complex, most audacious and most dangerous exploration mission in human history. That they didn't have the money to do it. That two hundred thousand people didn't actually apply. That, with all the good faith one can muster, I wouldn't classify it exactly as a scam – but that it seemed to be, at best, an amazingly hubristic fantasy: an absolute faith in the free market, in technology, in the media, in money, to be able to somehow, magically, do what thousands of highly qualified people in government agencies have so far not yet been able to do over decades of diligently trying, making slow headway through individually hard-won breakthroughs, working in relative anonymity pursuing their life's work. That he shouldn't look continually and fantastically to a theoretical future while his chance to be actually present in the privilege of human life passes him by. That he shouldn't give up on the hard work of making a life with the rest of us here on this horrendously messy, imperfect, unimaginably fragile and steadily warming Earth.

I flew across Australia from Sydney to Perth to tell him in person, thinking of how all the way.

Late in the day Josh and I were sitting across from each other sunk in very deep sofas, making us both look small, as the sun set and the room darkened around us. I asked Josh how he would feel if in the end Mars One didn't happen. If he made it all the way through to selection, but the mission just wasn't ever going to be real.

'Disappointed,' he said quietly after a long moment. 'Disappointed. But in the grand scheme of things it's already done.' Josh was quiet and reflective and drained, different from how he'd been every other time we'd spoken.

For someone like Josh, it is a quest for true purpose, for belonging, a burning wish to be exceptional. 'It's given me direction. It's given thousands of other people direction. I suppose why I have latched on to this so hard is that I once looked at the military as the be all and end all of things. Maybe we can move past the idea of having to defend ourselves from ourselves and be driven to explore.'

As we kept talking and some of the most insurmountable problems with the mission came up – the lack of money, the fact that the selection panel wasn't being made public, that there were no contracts with SpaceX – the more rational parts of Josh's thinking emerged. He is not a stupid person, not by any stretch. When I said that Chris Hadfield had serious reservations about Mars One, Josh said that he wasn't surprised and that other astronauts had expressed their scepticism, that he

The slopes of Mauna Kea, Big Island, Hawaii, photographed through the window of an astronomical observatory on the mountain.

knew about it. Especially that one astronaut in particular to whom he had always looked up shared the same outlook: Andy Thomas.

'He hates it,' Josh said. 'Absolutely hates it.'

Josh knew, on some level, that what Mars One was proposing was unlikely to come off. At least not in the time frame it set and not for the amount of money it said. But it was even that most minute, most remote chance it could actually work that kept Josh holding on to hope, the hope that brought him home from Europe to dedicate all his energies completely to Mars One. To keep trying to make it real.

'It's Joseph Campbell's *Hero With a Thousand Faces* type of thing,' he said, leaning forward to put his elbows on his knees. 'Except you stay in the hall of heroes, you don't return with the boon. You are sending it back, you're sharing it with the old world. But you're staying out there on the adventure calling others to come.

'That's why I'm willing to sign up to go one way.'

*

In 2019 Mars One declared bankruptcy.

The story of Mars One is not about a foolhardy attempt to achieve the impossible. It's a story about a credulous press. There would have been no story for me to report had there not been an enormous cohort of outlets willing to breathlessly hype an obvious pipe dream – which cost the people invested in it not only their money but their willingness to believe in a deluded, unachievable escape from their lives.

Mars One Ventures is reportedly more than $1.2 million in debt to creditors.

What disturbed me more than a bunch of people willing to exploit the naive enthusiasm of others, or the people themselves who were willing in theory to live out their days on an irradiated, lifeless rock of a planet for who knows what vague reasons, was the willingness of media outlets to uncritically champion what was from the start clearly never going to be a viable mission. This was by no means a Theranos – no one's life was ever put in danger – but was the product of the same brand of willingness to believe in a vainglorious promise of supposedly scientific triumphalism.

Some things sound too good to be true, because they are.

Everyone should demand a far higher standard of rigour from the people whose job is meant to be telling the public fact from fiction.

Our planet is boiling us all to death, and there is no escape plan to a planet B. That's the real story. 🖋

Digging Deeper

WATCHING

Ann Druyan, Steven Soter, Brannon Braga
Cosmos: A Spacetime Odyssey
2014
Cosmos: Possible Worlds
2020

Christian Frei
Space Tourists
2009

Werner Herzog, Clive Oppenheimer
Fireball: Visitors from Darker Worlds
2020

Glen Keane
Over the Moon
2020

Steven Leckart, Glen Zipper
Challenger: The Final Flight
2020

Ronald D. Moore, Matt Wolpert, Ben Nedivi
For All Mankind
2019–

Christopher Nolan
Interstellar
2014

Ridley Scott
The Martian
2015

David Sington, Heather Walsh
Mercury 13
2018

Stephen van Vuuren
In Saturn's Rings
2018

Denis Villeneuve
Arrival
2016

Robert Zemeckis
Contact
1997

READING: FICTION

Ted Chiang
Stories of Your Life and Others
Vintage, 2016 (USA) / Picador, 2020 (UK)

Arthur C. Clarke
Rendezvous with Rama
Harper Voyager 2020 (USA) /
Gateway, 2006 (UK)
The Fountains of Paradise
Orion, 2000 (USA) / Gateway, 2012 (UK)

James S.A. Corey
The Expanse series
Orbit, 2011–21

Michel Faber
The Book of Strange New Things
Hogarth, 2015 (USA) / Canongate, 2015 (UK)

Jaroslav Kalfař
Spaceman of Bohemia
Back Bay Books, 2017 (USA) /
Sceptre, 2018 (UK)

Ursula K. Le Guin
The Birthday of the World: And Other Stories
Harper Perennial, 2003 (USA) /
Gateway, 2010 (UK)

Liu Cixin
The Three-Body Problem
Tor Books, 2014 (USA) /
Head of Zeus, 2015 (UK)

Ian McDonald
Luna trilogy
Tor Books, 2016–20 (USA) /
Gollancz, 2016–20 (UK)

Kim Stanley Robinson
2312
Orbit, 2012

READING: NON-FICTION

Jim Al-Khalili (ed.)
*Aliens: Science Asks: Is There
Anyone Out There?*
Picador, 2017 (USA) / Profile, 2016 (UK)

Samantha Cristoforetti
Diary of an Apprentice Astronaut
The Experiment, 2021 (USA) /
Penguin, 2022 (UK)

Gordon L. Dillow
*Fire in the Sky: Cosmic Collisions, Killer
Asteroids, and the Race to Defend Earth*
Scribner, 2020

Matt Haig
The Humans
Simon and Schuster, 2014 (USA) /
Canongate, 2018 (UK)

Sarah Stewart Johnson
*The Sirens of Mars: Searching
for Life on Another World*
Crown, 2020 (USA) / Penguin, 2021 (UK)

Louisa Preston
*Goldilocks and the Water Bears:
The Search for Life in the Universe*
Bloomsbury Sigma, 2016

Mary Roach
*Packing for Mars: The Curious
Science of Life in Space*
Oneworld, 2011

Carl Sagan
Cosmos
Ballantine Books, 2013

Kip Thorne
The Science of Interstellar
Norton, 2014

Various authors
The Universe: A Travel Guide
Lonely Planet, 2019

David Whitehouse
*Space 2069: After Apollo, Back
to the Moon, to Mars ... and Beyond*
Icon Books, 2021

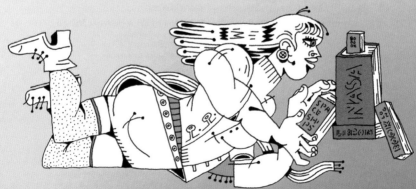

Playing the Universe

GIORGIO SANCRISTOFORO
Translated by Deborah Wassertzug

All of us have lingered to observe the moon, the stars, the planets or the Milky Way at one time or another, when weather conditions and the hour have permitted us to do so. But how many times have we *listened* to them? I'm not speaking about an acoustic sound but rather something more subtle and fleeting, which has stimulated the imagination of philosophers, scientists and musicians for millennia.

In the West, in fact, music and the universe have been intimately linked all the way back to antiquity – back to the time when Pythagoras, the philosopher and mathematician, discovered the relationship between geometry and sound that governs the physics of all musical instruments. Pythagoras was always steadfast in his conviction that numbers were a guide to interpreting the universe and that mathematics and geometry might explain everything – even music. Every musician, whether they are aware of it or not, has always had something to do with ratios, logarithms, frequency and functions.

Pythagoras' idea has echoed across time thanks to one of the best-known Platonic dialogues, *Timaeus*, in which the demiurge models the world-soul after a series of harmonic proportions. In the Middle Ages the concept solidified that mathematics, geometry, music and astronomy were fundamentally the same thing. Schools envisioned teaching these four subjects as what was called the quadrivium of liberal arts: mathematics studied numbers; geometry studied numbers in space; music studied numbers in time; and astronomy, numbers in time and space. The thought underlying the Platonic idea of the world was that the cosmos formed an immense musical composition in which the movement of the planets and the alternating constellations produced sounds imperceptible to the human ear and in perfect relation with one another. This composition was a testament to the perfection of

GIORGIO SANCRISTOFORO is a sound designer, artist and member of the 'Expert Artist' programme of the European Commission. He has created numerous software packages for experimental music (which have been used in over forty countries), produced documentaries and books on electronic music and has taught sound design and computer music for twenty years. He practises SciArt, creating works and installations based on nuclear physics and genomics.

the work of the divine. Human music could only be a reflection of this unattainable, majestic, heavenly music. The feelings and the moods that human music can stimulate were therefore explained with the philosophical properties of the planets: the belligerence of Mars, the lasciviousness of Venus, the vivacity of Mercury and so on.

Among the most important musical legacies from this period was the work of Hildegard von Bingen, the first person in Western music to be named as an actual composer. She wrote a voluminous collection of antiphons, responsorial chants and hymns – all inspired by the music of the spheres – known as the *Symphonia Armonie Celestium Revelationum*. It is one of the most significant musical collections from the 12th century and provides a key to the entire spiritual and mystical universe of this great female composer. In it we find an extraordinary attempt to reconcile the harmony of human beings with that of the heavens.

If the Middle Ages was fertile ground for planting *musica universalis*, the music of the spheres, the Renaissance and Baroque periods found the garden in full flower – in spite of the Copernican revolution having removed the Earth from the centre of the universe. The idea that the cosmos played a symphony remained firmly anchored in the thought of the great music theorists of the time, such as the Franciscan Gioseffo Zarlino, choirmaster of the Basilica of San Marco in Venice. He wrote one of the most influential musical treatises of the Renaissance, *The Harmonic Institutions*. In the fifth chapter Zarlino divides music into two parts: *animastic* music and *organic* music. While the latter defines harmony composed by various instruments, including voice, the former refers to the inaudible harmony that holds together the macrocosm and the microcosm. Zarlino,

with his precise heavenly and religious references, creates a space for discourse that will later be more thoroughly investigated and developed by such visionary minds of the Baroque as Robert Fludd, Athanasius Kircher and, most of all, Johannes Kepler. With Kepler the association between astronomy and music theory reaches its apex. He is not just a mere astronomer – his scientific contributions remain relevant today through his three famous laws of movement of the planets. In contrast to his predecessors he takes the trouble to demonstrate things through calculations, measurements and observations. In his famed *Harmonices Mundi*, Kepler modernises the concept of celestial music, adapting it to new astronomical discoveries.

Among the great musicians of the Renaissance and the Baroque who were inspired by the stars are Carlo Gesualdo da Venosa, Claudio Monteverdi and Antonio Vivaldi, while in the classical period we cannot forget Mozart. The music of the spheres is concealed within *The Magic Flute* behind an esoteric, symbolic veil. Everything in this opera has an initiatory meaning, including the characters of Sarastro and the Queen of the Night, symbolising the sun and moon. This should surprise no one, because the bond between the music of the spheres and Hermetic literature is amply reinforced in the works of Marsilio Ficino, Pico della Mirandola, Cornelius Agrippa and Athanasius Kircher, all authors who were welcomed and honoured by the Freemasons, of which Mozart was also an initiate.

Making a jump ahead in time, we arrive at the Romantic era, which we could in a certain sense consider a sort of pause during which man and his passions take the place of the heavens at the centre of philosophical and artistic thought. There is no lack of '*clair de lune*', of course, but we must wait until the 20th century to hear the music of the spheres in all its glory. At the beginning of last century, in fact, we have the famed orchestral suite *The Planets*, written by British composer Gustav Holst. The symphony has remained indelibly engraved on pop culture because American composer John Williams was strongly influenced by its first movement, dedicated to Mars, when writing his score for *Star Wars* – particularly the music that accompanies battle scenes involving the Empire and the Death Star.

In contrast to Holst's classical tendencies is *Pierrot Lunaire* by Arnold Schoenberg, based on a poem by Albert Giraud. The opera was Schoenberg's twenty-first work and is imbued with numerology. The ensemble and the conductor form a group of seven musicians. The use of seven – the number of planets known in antiquity – recurs in seven-note motifs throughout the piece.

Later, the post-war avant-garde was liberated from the burden of harmony – which had been undermined by late Romanticism and dissolved by atonal and twelve-tone music – and rediscovered the contrapuntal techniques of the Renaissance, which inspired a Pythagorean matrix that would foster a rapprochement between music and astronomy, at times conceptual and at times literal. The principal standard bearer of this new music of the cosmos was undoubtedly Karlheinz Stockhausen, who dedicated many of his works – such as *Sternklang* (Sound of Stars) and *Ylem* (The Breath of the Universe) – to astronomical themes. Meanwhile, the experimental music of John Cage transformed astronomical atlases into orchestra scores in his work *Atlas Eclipticalis*.

The technology of space voyages has finally made possible the realisation of the Renaissance dream of hearing the sounds of celestial bodies through the translation of scientific data into sound events.

In 2002 Terry Riley was commissioned by NASA to write *Sun Rings* for the Kronos Quartet, in which we hear fragments of recordings from the Voyager missions. The Voyager missions would also be used in the famous laserdisc of ambient music *Symphonies of the Planets*.

Electronic music has explored the universe since the tape music of the 1950s and continues to do so today with the involvement of the scientific community. Liberated from the limitations of traditional instruments, electronica incorporates interstellar whistles and electromagnetic waves in a rediscovered Pythagorean harmony in which human beings, just as in antiquity, find themselves facing the harmony of the spheres.

The next time you see a starry sky, remember to bring a pair of headphones!

A voyage through time ...

... and space

1

Wolfgang
Amadeus Mozart
The Magic Flute
1791

2

Terry Riley for
the Kronos
Quartet
Sun Rings
2019

3

Paul Hindemith
*Die Harmonie
der Welt*
1957

4

Karlheinz
Stockhausen
Sternklang
and *Ylem*
1971 and 1972

5

Gustav Holst
The Planets
1914–17

6

Arnold
Schoenberg
Pierrot Lunaire
1912

7

John Cage
Atlas Eclipticalis
1962

8

Gérard Grisey
Le Noir de l'Étoile
2004

9

Claudio
Monteverdi
*Al Lume delle
Stelle*
1619

10

Hildegard
of Bingen
*Symphonia
Armonie
Celestium
Revelationum*
1151–8

11

Carlo Gesualdo
da Venosa
*Mentre, Mia
Stella, Miri*
1594

12

Iancu Dumitrescu
Galaxy
1993
L'Orbite d'Ouranus
1990
*Meteors
and Pulsars*
1998

Stellar pop and electronica

You can listen to this playlist at:
open.spotify.com/user/iperborea

1

Fiorella Terenzi
Sidereal Breath
1991

2

Brian Eno
Under Stars
1983

3

Vangelis
Main Sequence
1995

4

Mike Oldfield
Supernova
1994

5

Louis and Bebe
Barron
*Forbidden
Planet: Main
Title Theme*
1976

6

Tangerine
Dream
*Birth of Liquid
Plejades*
1972

7

Timothy Drake
Jupiter
2016

8

Robert Schröder
*Galaxie Cygnus
A: Teil 1*
1982

9

Pink Floyd
*Welcome
to the Machine*
1975

10

The Orb
*Back Side
of the Moon*
1991

11

Isao Tomita
Space Fantasy
1978

12

William Basinski
4(E+D)4(ER=EPR)
2019

Recording the Universe

GIORGIO SANCRISTOFORO
Translated by Deborah Wassertzug

On 11 February 2016 the scientists of the Laser Interferometer Gravitational Wave Observatory (LIGO) at Hanford in the US state of Washington announced to the world that they had detected gravitational waves for the very first time. Two black holes, some thirty times more massive than the sun, had fused into one a billion and a half years ago, generating a disruption in space–time that spread across the universe like the ripples on a lake when you throw a stone into the water.

In September 2019 a team of researchers from LIGO taking another look at the data discovered other important details about this event. The new black hole had vibrated in a way that was not very different from a bell. In fact, the signal had *overtones*, meaning it had a sort of *timbre*.

That same year, American composer William Basinski used the data from this experiment to create the EP *On Time Out of Time*, produced for the installation ER=EPR by the Belarusian artist Evelina Domnitch.

Let's talk to the scientist and researcher Maximiliano Isi of the Massachusetts Institute of Technology, who heads the team that discovered the 'sound' of the black holes.

What are gravitational waves, and how do you capture them?
Gravitational waves are disturbances in space–time generated by violent collisions or by large masses in movement through the cosmos in a way that is similar to how soundwaves travel through air. The effect of these waves is to modify distances by stretching or contracting space. We can observe gravitational waves by using instruments called interferometers, which can measure microscopic variations in distance. These instruments are quite similar to gigantic microphones. We shoot a high-intensity laser and divide it into two perpendicular beams in two tunnels that are of identical length, four kilometres. At the end of each tunnel a mirror reflects the laser beam back. In the presence of a gravitational wave, the distance of one of the two tunnels varies a fractional amount, and this helps us to observe a contraction or expansion of space. It's a very precise system that lets us measure variations of distance thousands of times smaller than the nucleus of an atom. It allows us to analyse the characteristics of very distant objects, and it confirms our knowledge about the nature of space–time. So far we have observed about fifty large mass collisions in space, such as black holes and neutron stars.

As a musician, what struck me about your article was that the collision of the two black holes created what you call 'overtones'. In other words, when these two objects fused into one the vibrations they generated are quite similar to those produced by a bell. How are these waves transmitted?

Gravitational waves are disturbances in the curvature of space and time. They move at the speed of light and traverse matter practically undisturbed, which makes them hard to observe. We can imagine space–time as a membrane that is perfectly elastic: it can fold and twist in response to the effects of matter and energy – this is the principal lesson of Einstein's general theory of relativity. Space–time tells matter how to move, but matter tells space–time how to curve. Since gravity is nothing more than the manifestation of the curvature of space–time, you can think of gravitational waves as waves within the gravitational field, just as light is a wave in the electromagnetic field.

These days astrophysics and astronomy often use sound to disseminate scientific information. Science is a world principally composed of graphs and equations. Do you think that analogies that use sound are valid in aiding understanding?

In the case of gravitational waves, the comparison with sound is apt – not only because gravitational waves are waves but also because the signals that we receive on Earth, by pure coincidence, are in the range of frequencies that humans can hear (between ten and a thousand Hertz). So we can easily take our data and listen to it without even needing to transpose the numbers. Another example of how well the analogy of sound fits is in the way our instruments work. They are basically huge microphones that capture vibrations from every direction. For this reason we need to use a number of them to triangulate the source and understand where the signal is coming from. In other branches of science the connection with sound is perhaps less direct, but it is often useful to try to listen in order to make complex concepts more intuitive and communicable.

Recent scientific discoveries have confirmed ancient beliefs which state that everything we are and everything that surrounds us is composed of vibrations. It seems increasingly apparent that the intrinsic properties of objects, of matter, are less important than the relationships they create between them. What do you think?

Yes, perhaps. It is true that theories of modern physics explain the world in terms of fields vibrating in relation to one another. But is that the ultimate essence of things? It is a possibility – it is, among other things, what the well-known string theory suggests.

I'll wrap up with a philosophical question. What if our innate predisposition for music were proof of a sort of celestial harmony, and we are actually living in a great symphony?

As you've correctly noted, vibrations, resonance and harmony are phenomena that are omnipresent at every level, from the most abstract to the most banal. Obviously the origin of our innate predisposition for music has to do with the importance of sound for communication and survival. We are also profoundly social animals who need to transmit emotions and meaning. This also determines the way in which we perceive and process the world: we draw pictures, play music, tell stories. I don't know what the fundamental nature of reality might be, but whatever it is, once we have found it (if we ever do) we will make sense of it through images, music and stories.

A symphony? Why not?

SITUATIONS VACANT

AAASTROBOTANIST WANTED

Are you green fingered? Flex them in the most extreme conditions and save our Mars colonies from a life of freeze-dried slop and packaged food! **Food and lodging** – and **gratitude** – **provided**. Astrobotany degree required. Seeds not supplied.

EXPERT IN OUTER SPACE LAW

Company in the mineral extraction sector seeks legal counsel familiar with finding loopholes to help us exploit resources on the moon, Mars and various asteroids. Tax advice welcome. **Low pay** but the successful candidate will have the opportunity to influence space law for generations to come. (**Good for CV**)

XENOLINGUIST-TRANSLATOR/ INTERPRETER

Seeking linguists, translators and interpreters with great imaginations to study communications captured by the FAST radio telescope in China from a nearby galaxy. Appears to involve an agglutinative and ergative language. Knowledge of Basque preferred. **Send CV urgently** as tone of message sounds threatening.

DELIVERY DRIVERS IN OUTER SPACE!

Seeking driver for deliveries of food, drink and various objects to the International Space Station (ISS). **Licensed rocket owners** preferred.

★★★★

999-103 033 989

SPACE ELEVATOR OPERATOR

Seeking operator for space elevator between Earth and low Earth orbit. Duties include loading and unloading goods, accompanying passengers, DJ upon request (each trip lasts three days up, three days down). Must be presentable and entertaining. Uniform provided by **Otis-Schindler Space Elevator Company**. First aid and firefighting qualifications required.

WANTED: NEUTRINO HUNTER

Do you have a passion for the invisible? Not afraid of lost causes? **Become a neutrino hunter**! You are unlikely to capture or observe one, but satisfaction guaranteed in the event of success. **Excellent pay**. Perfect for an agoraphobic.

SPACE REFUSE COLLECTORS SOUGHT

We are seeking ten space sweepers to clear Earth's orbit of space probe and satellite detritus. **Flexible hours**. Not suitable for vertigo sufferers. Spectacular views.

SPACE ARCHITECTURE STUDIO

Become a pioneer on a new frontier of architecture! You will design the interiors of orbiting structures and planetary bases, create spaceports on Earth and design space yachts, experimenting with materials, differing gravities, alternative energy sources and robotics. **Remote work possible**. HQ in Luxembourg.

★★CALL NOW!!!★★

BALLOON PILOT

Seeking pilots for the latest Venusian balloons helping scientists study the inferno below. **60km** above the surface of Venus. Uniform: shorts and T-shirts. Atmospheric pressure similar to Earth, and air is breathable (even if it smells like rotten eggs). **Experience essential** – winds can be strong and falling to the planet's surface means certain death.

Graphic design and art direction: Tomo Tomo and Pietro Buffa

Photography: Francesco Merlini, Samuele Pellecchia, Raffaele Petralla, Massimo Sciacca, Gaia Squarci, Scott Typaldos

Photographic content curated by Prospekt Photographers

Illustrations: Edoardo Massa

Infographics and cartography: Pietro Buffa

Managing editor (English-language edition): Simon Smith

Thanks to: Viola Angelantoni, Matteo De Giuli, ESA, Peppo Gavazzi, Francesco Guglieri, Monica Malatesta, Emanuele Menietti, Andrea Morstabilini, NASA, Dario Scovacricchi, Kelly and Zach Weinersmith

http://europaeditions.com/thepassenger
http://europaeditions.co.uk/thepassenger
#ThePassengerMag

The Passenger – Space © Iperborea S.r.l., Milan, and Europa Editions, 2022

Translators: Dutch – Laryssa Aijal; Italian – Lucy Rand ('The Underground Universe'), Alan Thawley ('Looks Lovely, But I Wouldn't Live There', 'A Brief Guide to Achieving Orbit on a Budget', 'Reaching for the Stars', infographics, editorial, sidebars, standfirsts, captions), Deborah Wassertzug ('Space Barons', 'Living in Space', 'Playing the Universe', 'Recording the Universe')

Translations © Iperborea S.r.l., Milan, and Europa Editions, 2022

ISBN 9781787704190

Printed on Munken Pure thanks to the support of Arctic Paper

Printed by ELCOGRAF S.p.A., Verona, Italy

Secondary School

Assemblies

- for Busy Teachers -

Vol 2

CheckPoint
Press

SECONDARY SCHOOL ASSEMBLIES FOR BUSY TEACHERS Vol 2

ISBN-13: 978-0-906628-70-3

Published by CheckPoint Press, Ireland

CheckPoint Press
Books With Something to Say..

email: editor@checkpointpress.com

www.checkpointpress.com

Cover and illustrations by Julie Hayes

SECONDARY SCHOOL ASSEMBLIES

FOR BUSY TEACHERS

Vol III

INTRODUCTION

All but one of of the assemblies in here are ready to go!

We know that those lucky enough to be presenting an assembly on a Monday morning haven't always had the time to plan and prepare a masterpiece. This is where this book comes in.

We would, of course, recommend reading the material through first, because a little practice does wonders for any delivery, and because some of the assemblies include interactive aspects.

Do read the sections *in italics* before beginning as these are instructions, guidelines or suggestions for the presenter.

But if the worst comes to the worst, and you only have time to briefly survey the contents, you can still grab this little book, go into your assembly hall and simply read one out.

However, if you are in this much of a rush, it may be better to skip the 'audience interaction' bits, or make the optional questions **(as listed 'Q' in bold)** rhetorical, because once your students join in you really need to know where you are going!

While being predominantly secular, all of these assemblies have a celebratory and enquiring attitude to life, reinforcing positive values and attitudes to one's self and others.

Enjoy!

Contents.

HOW TO STRUCTURE AN ASSEMBLY

Speaking in public is not like other forms of communication. It can be a very nerve-wracking experience facing a host of expectant, eager faces - some of whom no doubt, will find great amusement in any errors or blunders made by the presenter.

Being aware of what makes public speaking different from other forms of communication will help us understand what we need to do in order to give a great assembly.

Speaking to large groups of people on formal occasions presents a unique set of challenges which need to be dealt with differently.

1. How an Assembly differs from a conversation:

    It is harder to command your audience's attention than when talking with two or three people because, unlike in an informal chat, the people who are listening to you do not expect to have to contribute; they may switch off!

    In a small, informal group, you are rarely expected to be the expert, or to come out with some pearl of wisdom. But in a speaking engagement (which is what an assembly really is) you are supposed to be worth listening to – that's why you have been given the privilege of the stage.

2. An assembly is *not* the same as simply handing out a document or narrating a slide show:

It is harder to command your audience's attention than if you had given them something to read, or look at, because, unlike reading a document or a book, the people listening to you cannot go back and check something they didn't hear properly or understand. So if they don't 'get it' loud and clear the first time, you will lose them.

So you really are up against it a bit. The assembled students can just sit there in silent judgement of you; they can ignore you; they can secretly laugh at you; or they can fail to pay you any attention whatsoever.

And you may not even know it!

So mostly, this book is about interesting content. We have written what we believe is some compelling, topical material. But we must always remember that form and content are interlinked.

Here are a few pointers for a great presentation.

PERFORMING TIPS

Think of yourself as 'a performer'. It helps. This is not a natural easy-going conversation. Even if you make it seem like one. It's a performance.

Don't stand behind a table or a lectern all the time as it creates a psychological barrier between the speaker and the audience. Move away from the lectern when making important points or engaging with individual audience members.

Audience interaction can be a double-edged sword. Once you invite audience participation, the power-base of the presentation shifts as the students realise they can take part. It is important that you maintain control of the direction of the assembly, so be careful if fielding 'open' questions.

Some of the assemblies lend themselves well to the use of props, visual aids and power points. But take care that these aids support, and do not distract from the central message. Keep the assembly interesting by keeping it short, sharp and succinct.

Movement

Avoid being completely predictable in your movement: don't stay in the same place the whole time, especially if this involves being stuck behind a table or a lectern.

Make clear decisive movements: a clear decisive movement in almost any direction will make your audience sit up and take notice. Try to do these at significant points in your delivery. It will look as though you are moved (literally) by your own words. However, don't keep repeating the same movement such as pacing backwards and forwards, as this quickly becomes obvious and/or distracting.

Look at Them

Make eye contact as much as possible and scan the whole room. Don't just look at the front or the back, the right or the left, or at one or two friendly faces. Systematically scan the room, and give the audience the feeling you are talking to each of them *personally*.

Passion

Always, always, always speak as if you are deeply serious about your message. Make it seem that it means a lot to you. If it doesn't mean much to *you*, then why should *they* listen?

An overly-casual attitude undermines you, and undermines the material you are presenting. It would *not* be helpful for example, to open with a casual comment like; *"I knew I had to do assembly today so I just looked on the internet to find something to talk about."*

Stories

A great way to deal with an assembly is to have a story at its heart. People love listening to anecdotes, especially well-constructed ones, and it gives you a central theme to hang everything else around.

Structure

Follow a clearly defined route. We have structured the assemblies in this book as follows:

- Opening Hook
- Overview
- One/Two/Three Main Points
- Summary
- Closing

1. Opening – start with a 'hook' - something to make them sit up and take notice.

> As a public speaker, you should always be looking for ways to make your audience listen; to catch and hold their attention. Telling them that something is unique about any particular assembly is a great way to start. Other classic openings are the use of a rhetorical question; an amazing fact; an interesting story or a famous quotation.

2. Introduction and Overview – Tell them generally what you are going to tell them, without giving it all away.

3. The Body of the Assembly – One, Two or Three Main Points with clear links between them. Links are very important. They are like signposts for your audience which bind the presentation together and give it cohesion and meaning.

4. Summary – Tell them again briefly, what you've already told them. Hit all the main points again.

5. Closing – A mirror of the Opening; something with a bit of drama, humour or punch!

Finally, don't forget to thank the students for their time and attention.

Structure

Follow a logical, fluid route. We have outlined the assemblies in this book as having:

- Opening Hook
- Overview
- One, Two, Three Main Points
- Summary
- Close

1. Opening - Start with a "hook" - something to make them sit up and take notice.

As a public speaker, you should aim to be 'looking for ways to grab your audience's attention, to grab and hold their attention'. Telling them that something is unique is one way. Jokes/anecdotes, ambly is a great way to start. Other classic openings are the use of a rhetorical question, an amusing or thought-provoking story or a humorous question.

2. Introduction and Overview - Tell them generally what you are going to tell them, without giving it all away.

3. The Body of the Assembly - One, Two or Three Main Points with clear links between them. Links are very important. They are like signposts for your audience which bind the presentation together and give it cohesion and meaning.

4. Summary - Tell them again in brief, what you've already told them. Recall the main points again.

5. Closing - A mirror of the Opening, something that will with what rounds it off.

Finally, don't forget to thank the students for their time and attention.

10

I

TERRY FOX

AIM: To encourage students to always do their very best, regardless of the circumstances.

Time: 10 - 15 mins.

HOOK

Have you ever run for the bus or competed on sports day or moaned about having to run a 3 mile cross country course? Hard work, isn't it?

Can you imagine how fit you would have to be to run 3,339 miles – which is just over one thousand times your average cross country course? Now imagine running this distance when you only have one leg. Also, you are dying of cancer. This is what the person I want to talk to you about today, managed to do.

OVERVIEW

Terry Fox was a Canadian boy whose favourite sport was basketball. Unfortunately, he was too short and just not good enough to make his high school basketball team. After only a few practises with the team, his coach suggested he might be better off playing another sport. "How about wrestling?" the coach suggested.

At school, Terry was very introverted and lived in fear of his teachers asking him questions during lessons. He was so small that, in Year 8, his feet didn't touch the floor when he sat at his desk. He was passionate about basketball so, despite his coach's discouraging words, he left for school early every morning so that he could practise before school. During his first season, he worked as hard as he could even though he only got one minute on court during the entire season. Can you imagine doing this?

Terry was tenacious and determined. By Year 10, he was in the starting line-up of the team and by Year 12, he won athlete of the year. His determination earned him a place on a highly competitive university team – the first stepping stone to his dream of playing professional basketball.

MAIN POINT 1: DEALING WITH UNEXPECTED SETBACKS AND CHALLENGES.

During his first year of university, he made a stupid mistake. He let himself be distracted when driving, crashed his car and injured his knee. Six months later, he was still suffering from serious knee pain. He was afraid of being kept off the basketball court so waited until the end of his season before finally seeking medical help. Terry went in to the hospital with a knee injury but walked out with the worst possible diagnosis: cancer that would require his leg to be amputated above the knee, sixteen months of chemotherapy ahead of him and only a 50% chance of survival. He was grateful though as only two years earlier, his survival chances would only have been 15%. In other words, for every 100 people who developed this type of cancer, only 15 would survive. When Terry was diagnosed, 50 out of every hundred survived. His odds were still not good.

During his long period of chemotherapy, Terry watched many other patients die. Doctors attributed his survival to his determination to overcome the cancer. Within three weeks of his amputation, he was up and walking and soon joined his father on the golf course. He went on to play wheelchair basketball and won many awards.

Motivated by the story of the first amputee to complete the New York Marathon, Terry ran a marathon in British Columbia. He finished last and was 10 minutes behind the next nearest runner: however participants and fellow runners were moved to tears by his accomplishment.

MAIN POINT 2: TURNING DEFEAT INTO SUCCESS.

Terry was aware that, thanks to medical research and advances, his chances of survival had improved from 15% to 50% in just two years. This inspired him to want to give something back. This is what he did:

The Terry Fox Marathon of Hope began on 12th April 1980. Terry's aim was to run the breadth of Canada – a distance of about 5,000 miles. He set out from the east coast with a support team that consisted of only a campervan and his best friend, Doug.

A few big companies such as Ford, Imperial Oil and Adidas offered Terry some strings-free support that he gladly accepted. Other companies wanted him to endorse their products in exchange for support. Terry turned these offers down and insisted that no-one should benefit from his run apart from cancer research.

Despite efforts to raise awareness of his run, as Terry began running through the eastern provinces of Canada, no-one really noticed. He was so depressed about this that he took his anger and frustration out on Doug and after only a few provinces (remember, there are ten), they weren't even speaking to each other anymore. Doug's brother had to join them to keep the Marathon of Hope moving.

Finally, with the help of a few key individuals, the Marathon of Hope picked up momentum. The owner of a major hotel chain, whose son had recently died of cancer, picked up on Terry's quest. He helped to provide food and accommodation, pledged $2 a mile to the cause and then persuaded nearly 1,000 other corporations to do the same.

Following Terry's arrival in Ottawa (the capital of Canada) on Canada Day, his charity raising became high profile with many politicians, sports personalities and celebrities supporting and promoting his cause. Hope was high and by now, crowds lined the roads, cheering as he ran by.

By 1st September 1980, Terry had reached Thunder Bay and was coughing badly as he ran. The crowds turned out to cheer him on and, not wanting to let anyone down, he continued to run for a few more miles. Finally, he had to be taken to hospital to face his worst fear: he had run his last mile.

143 days and 3,339 miles later (and doesn't that make cross-country look easy), his determination, tenacity and contagious enthusiasm had raised $1.7 million for cancer research.

What Terry didn't know was that, even though his death was imminent, this total would continue to grow and grow. Before he died, his goal of raising $1 for every Canadian ($24.1 million at the time) was achieved.

At last count (2010), more than $550 million dollars (£324 million) has been raised in his name. Every year, millions of Canadian school children (and adults) complete the Terry Fox Run to raise money for cancer research.

SUMMARY

One dream, one person, millions inspired. One scrawny high school boy decided life wouldn't beat him and kept trying. When he got knocked back, he came back harder. He wasn't rich; he wasn't poor. He wasn't anything exceptional to begin with – just like most of us.

So what made the difference? He was resilient – he didn't give up and he worked hard, and then harder again, to get what he wanted. He worked for himself and he worked for others.

CLOSING

Each of us, no matter how ordinary we may seem, has the capability within us to do great things.

Remember: one person, 143 days and 5,373 kilometres (3,339 miles), $1.7 million that has since grown to over $550 million. What great thing could you do? What great thing could the person next to you achieve?

If you want to follow Terry's dream of finding a cure for cancer, then remember that cancer affects 1 in 4 people so you will know at least one person who suffers from cancer during your life. You can go to Cancer UK online and help to fundraise. There are runs and other events all over the country and they always need more support.

Thank you for listening.

Notes:

2

NUMBERS

AIM: Bringing the concept of big numbers such as 'millions' and 'trillions' into context along with a general understanding of the importance of learning mathematics.

Time 10 to 15 minutes

HOOK

How much is a trillion pounds?

Britain's national debt was 1 trillion pounds in February 2013. We owe £17,869 (pounds) for every man, woman and child in the country. That is more than £38,344 for every person in employment. Every household will pay £1,919 this year just to cover the *interest* on our debt.

OVERVIEW

But how much is a trillion really?

One trillion is a lot. There can be some confusion over how much a trillion is as the English and Americans used to use different values for a trillion. We will assume a trillion is a 1 with nine zeros the original American definition that has been adopted by most countries (One thousand million)

Well, if you had a trillion pounds in front of you and were told you could keep all the pounds you can count, how much would you get if you had all the time in the world to count them? Let us assume you could count 2 pounds every second (which is pretty quick). In one minute you would have £120.

In 1 hour you would have £7,200.

In 1 day you would have £172,800.

You would make your first million after about 6 days – provided you didn't sleep, eat, or take any toilet breaks!

It would take you 15, 855 years – to count one trillion pounds.

One trillion is a lot.

MAIN POINT 1: NUMBERS ARE INTERESTING.

Some interesting number facts.

Fact number 1: The richest person in the world is Carlos Slim, owner of America Mobil Telecoms Group. He has 73 billion Dollars (£45, 742, 214,424.46) as of February 14th 2013. Bill Gates is the second richest. After giving away most of his money to charity, he still has $67 billion. To catch up with Bill Gates would take you 387,731 days or 1,062 years of counting. To be as rich as Carlos, you would have to count for 422, 453 days or 1,157 years.

Fact number 2: Google is a well known search engine but what does the word Google mean? Well it is a misspelling of the word googol which was the word chosen by a young child of the mathematician who named the number. It is the number '1' followed by 100 zeros.

Fact number 3: A Googleplex is a '1' followed by a Google of zeros which is a number so big it is impossible to write because there is not enough space in the universe.

Fact number 4: Are you beautiful? Is the person sitting next to you? Is your teacher absolutely gorgeous? Well if they are, it is all down to maths and the golden ratio.

Ratio and proportion are what produce beauty. If the distances between the features on your face are in the ratio of 1 to 1.618 then you are stunning. This is the exact ratio you find when you analyse the Mona Lisa. In fact, the chap who painted it, Leonardo da Vinci, was pretty skilled with numbers along with designing things like the helicopter and painting a few pictures. The belief in this ratio is now so strong that Chinese women applying to be flight attendants have their faces measured as part of the application process and those that don't add up don't get the job.

Fact number 5: 'Fractals' are some of the most beautiful patterns in the world. They come from numbers. *(Possibly show some fractals on a power point presentation).* A fractal is a shape whose small parts are identical or nearly identical to the whole shape – like a snowflake, a fern or a tree. In fact all living things seem to have fractal properties. So without them perhaps life would not exist.

Fact number 6: Odd facts about circles.
A sphere has two sides. A spider, for example, could crawl around on the inside or the outside of a sphere.

Among all shapes with the same perimeter a circle has the largest area.

Among all shapes with the same area a circle has the shortest perimeter.
(If practical, the presenter can involve the assembly in these Maths party tricks, allowing enough time for the students to keep up)

Trick 1

Step 1: Ask everyone to think of a number below 10.
Step 2: Double the number you thought of.
Step 3: Add 6 to the result.
Step 4: Halve the answer (that is, divide by 2).
Step 5: Take away the number you first thought of.
You should all have the answer 3!

Trick 2

Step 1: Think of any number
Step 2: Subtract the number you first thought of
Step 3: Multiply the result by 3
Step 4: Add 12 to the result
Step 5: Divide the answer by3
Step 6: Add 5 to your answer
Step 7: Take away the number you first thought of from the answer
You should all be thinking of 8!

These are not really 'tricks' they are just ways of using a sequence of mathematical operations to get the same answer. All developed by clever mathematicians.

MAIN POINT 2: NUMBERS *ARE* IMPORTANT.

So how important are numbers? Most people think that literacy is far more important than numeracy. It turns out that this may be completely wrong. A lack of numeracy skills, even if you're really good at reading and writing, can really mess up your life. For example, boys with poor maths ability risk being excluded from school and as men, are far more likely to be arrested and end up in prison. People whose maths aren't up to scratch tend to earn less (by approximately £2,000 a year) and may even suffer from physical or mental health issues.

In a survey conducted on the subject, men and women with poor numeracy skills—even those with good literacy skills—were in worse jobs, were more at risk of depression, and had little interest in politics. Even though women have been allowed to vote for nearly a hundred years, women with poor maths skills today are much less likely to voice their opinions in elections, and as we all know voting is probably one of the most important things you can do.

For men and for women numeracy skills are likely to decline if they are not used in employment. This forms a vicious circle where skills are not used, then lost, and employability declines rapidly. In today's society there are far fewer jobs around that do *not* require mathematical ability.

It turns out that poor numeracy skills can even shorten your life. Poor numeracy skills and a lack of understanding of personal finances for example, can lead to depression which in turn can lead on to suicide. Suicide is the most common cause of death in men under 35 in the UK.

SUMMARY

Well you may not like maths lessons much but we have heard today how important it is to be numerate. So pay attention the next time you are factorising an equation or doing a SURD. We have seen that maths can be fascinating, even vital to a happy and productive life, and how competence in numeracy is arguably THE most important skill set you can learn in school.

Thank you for listening.

3

BEING 'COOL'
AND KNOWING 'COOL'

AIM: To develop awareness amongst students of the transitory nature of passing fads, fashions and popularity, and the value of discernment when assessing what is 'cool' and what is not.

Time: 10 to 15 minutes

HOOK

Have you ever watched the television series Top Gear with that nice polite man, Jeremy Clarkson? If you haven't, it is sort of about motoring but with a lot of silliness and humour thrown in. It is one of the most popular television programmes in the world. Sometimes part of the show involves a 'Cool Wall' where the presenters place cars in different positions depending on how cool the audience feels the cars are. But what do we really mean by the word 'cool'?

OVERVIEW

Are there any pop stars, sports personalities, actors or scientists whom you think are cool?

(Pick three students and ask them the following question. Then put the names of the celebrities they choose and the 'cool' attributes of each as examples for the student body to see.)

Q: Can you name three people whom you think are very cool and tell me why?

So, based on these examples would Shakespeare be considered cool? In the view of the English department, he is probably the coolest guy ever. But in the view of Year 10 students he is probably the *least* cool person in the history of the world.

MAIN POINT 1: 'COOLNESS' IS RELATIVE.

I would like to tell you about two people, first of all King Canute.

Canute was a Viking King whose followers thought he was the coolest person ever. He was the first King to successfully rule over the whole of England. This left England free from internal and external strife.

Because Canute also ruled the Viking homelands he was able to protect England against attacks, maintaining twenty years of peace. This meant trade and social stability became established. He had a sense of fairness and justice that developed English law into a fairer system than that which went before.

He was also a Christian who looked after the population well: building churches and giving gifts to make amends for the past wrongs of previous Viking invaders.

He became so popular that his followers felt he was as cool as anyone can possibly be. In fact, they thought he was so cool that he could stop the tide from coming in – if he wanted to. Canute, however, was a smart politician who knew his limits and was confident enough in his own coolness that he wanted people to admire him for who he really was and not for some fictitious powers he didn't actually possess. To demonstrate this to his people, he had his servants transport his throne – not an easy thing to do – to the edge of the sea as the tide was coming in. He then sat on it and commanded the tide to stop. Of course, all that happened was that he got very, very wet as the tide came in over his throne, threatening to drown King Canute and all his entourage. At the very last minute, he moved back to the shore and explained to his followers that although he could do many great things as a man, he was nothing in the face of God's power. In the eyes of his followers, this made Canute even cooler than before. **Q: So what do you think? Was King Canute cool or what?**

Here's another example: 'Chauvinism' is generally understood today as a belief that your nation, group or sex is superior to everyone else's. So, are YOU a chauvinist? Do you know any chauvinists? Are chauvinists cool?

The word chauvinism comes from someone who took 'coolness' to such extremes that eventually it made him very uncool. Nicholas Chauvin was an accomplished soldier and a fanatically devoted follower of Napoleon. This made him really 'cool' while Napoleon was popular and in charge, but after Napoleon was discredited, Chauvin was suddenly *very* uncool – so much so that his name has entered the language as a description of 'uncoolness'.

An absurd loyalty like Chauvin's can be very dangerous – especially when it focuses on one's own group. Chauvinism can lead to denigration of others, feeling other groups are inferior. This can lead to bullying, gang culture and, in extreme cases, even genocide. "We are the best" (we think) so therefore, the rest of you are somehow less than us – so we have the right to abuse you or treat you as 'the enemy'.

So, how can something that makes you cool at one time make you so uncool at another? It seems that 'coolness' is relative to the circumstances – and relative to how we view things – right? Someone may be cool in your opinion but it may be wise to think twice before beginning to hero-worship them and believing everything they say.

Ask yourself, are the coolest kids in school the ones who are always in trouble, or are they the ones who always get top marks in exams? Is it cool to wear your tie too short, your pants too low, your sleeves rolled up.. or would this be the height of uncoolness? I remember a time (the 70's) when it was 'cool' for boys to wear long hair, high heels and flared trousers. Would that be cool now? No, I didn't think so!

Let's test your ability to detect the coolest leader. Imagine it is time to elect a world leader, and your vote counts. Here are the facts about the three leading candidates:

Candidate A: Associates with crooked politicians and consults with astrologers. He's had two mistresses. He also chain smokes and drinks eight to ten vodka martini's a day.

Candidate B: He was kicked out of office twice, smokes all the time, sleeps until noon, used drugs in college and drinks a quart of whiskey every evening.

Candidate C: He is a decorated war hero. He's a vegetarian, doesn't smoke, drinks an occasional beer and hasn't had any extra-marital affairs.

Q: Which of these candidates would be your choice? *[List the candidates again and ask for a show of hands].*

Let's see who you chose.

If you chose Candidate A, you voted for Franklin D. Roosevelt – an American President who was famous for his successful economic policies and helping to create the United Nations.

If you chose Candidate B, you voted for Winston Churchill – most famous for helping to win the Second World War.

If you chose Candidate C, you voted for Adolf Hitler – most famous for murdering millions of innocent people in the largest genocide ever known.

So.... now, who do you think is the coolest?

Hitler had a very persuasive personality and was hero-worshipped by the Nazis - in other words the Nazis thought he was cool. But a lot of Hitler's popularity came from manipulating the emotions of the German people. He did this by telling the Germans they were superior to everyone else, and that this somehow gave them the right to bully and condemn others. As a result, millions upon millions of people died.

So, being cool can be a good thing or a bad thing – depending on the reasons and the results. When David Beckham gets us all joining in his 'sports for kid's' campaigns, this is a good coolness. But when being cool means lots of people are left out, bullied or victimised, then 'being cool' is definitely a bad thing.

Think of someone you consider 'cool' in your class. Do they do things you want to do, do they have the things you want to have? Are their hair, clothes and shoes just the way you want yours to be? Are they good at sport? Do they seem to know everything about music, the best places to shop, and the movies everyone should see? Have they got the best boyfriend or girlfriend? And what would *you* do to be their friend?

Hitler's admirers were prepared to kill for him. I'm pretty sure none of us want to go that far – but how far *would* we go? When does admiration turn into blind hero worship? And when does that become dangerous?

Well, Jeremy Clarkson is certainly well-known, popular and entertaining. But whether he is really 'cool' or not we will leave you to decide.

Thank you for listening.

"Gone, but not forgotten"

ACHIEVEMENT

AIM: This assembly considers the achievements of key people in history and asks students to reflect on the difference that these people have made to our lives. It also asks students to reflect on themselves – where their own strengths lie and the qualities they have – and what they might achieve in their own lives and the difference this could make to others.

Optional Props: signed photograph or signature; baby handprint and the mark of a cross, set of footprints.

Time: 10 minutes approximately.

HOOK

I would like you to think for a moment about what the following have in common *(the presenter may wish to show pictures at this point)*. Here is my list – a baby handprint, a set of footprints, a cross and a signed photograph.

Q: Can anyone tell me what the link is between these? *(Field suggestions from the student body)*

A: These are all ways in which someone has 'left their mark' or 'made their mark'. We leave footprints in the snow or sand where we have walked. Parents will often take a handprint of their baby as a memento. People would make a cross as their mark in years gone by if they could not read or write, and today, adults identify themselves with an individual signature.

OVERVIEW:

So in today's assembly I want you to think a little about the ways in which people through history have left or made their mark. We will consider two examples of famous people who lived at a similar time in the twenty first century and look at their achievements. Through this, I want you to consider what qualities these people had and reflect on your own life and ambitions – what you hope to achieve and what will people remember YOU for.

MAIN POINT 1: THE IMPORTANCE OF CONTRIBUTING.

Can you put your hand up if you have watched television in the last twenty four hours? Have you sent an email or used the internet? Has anyone turned on a light, taken some food from the refrigerator, had a lift in a car or boiled a kettle?

Where would we be without these key inventions?

Life would be very different and certainly not as comfortable if we didn't have electricity, vehicles, computers and the internet. All through history, whether it is through inventions or in other ways, special people have made their mark or done something extraordinary that has changed the lives of others. Let's think of a few examples. As I read each one, you might want to reflect on where we would be without these achievements and what personal qualities you think these people had.
(The presenter may wish to make the following list interactive to see if students can give the answer for each person. Alternatively, the list can just be read out)

- Tenzing Norgay and Edmund Hilary – the first people to reach the summit of Mount Everest,
- Roald Amundsen – the first person to reach the South Pole,
- John Berners Lee, the person who invented the world wide web.
- John Logie Baird who invented television.
- Florence Nightingale the pioneer of nursing.
- Abraham Lincoln the 16th president of the USA, who united the states and abolished slavery,
- Thomas Edison the inventor of the light bulb.
- Karl Benz who invented the first car.
- Amelia Earhart the first women to fly solo across the Atlantic Ocean.
- Charles Babbage who invented the first computer.
- The Wright Brothers' first flight.
- Elizabeth Fry the prison reformer.
- Alexander Fleming who discovered penicillin.
- Marie Curie the first female to win the Nobel Prize.

There are many more people we could list and I am sure you will have examples of your own. These people dedicated their lives to a cause which made changes to the lives of others in the years that followed.

MAIN POINT 2: TWO NOTABLE EXAMPLES.

I want to spend a little time now finding out a bit more about two people who very clearly left their mark on the twentieth century – albeit in different ways. They both lived in the USA, were born within a year of each other and made history in the 1960s. The words that they spoke are famous and can be instantly recognised. They pushed boundaries even in the face of opposition; they did not accept failure and they made history. One died whilst fighting for his cause whilst the other, just one year later, took probably the most famous step in history. They were both 39 years old. Any ideas yet?

Let's see if you know who they are? Here are some of the words they spoke. *(Read out the quote or play a short clip of the moon landing if available)*

"That's one small step for man, one giant leap for mankind."

These words were heard by an estimated 600 million people watching the 1969 moon landing on TV. It is perhaps the most famous phrase spoken in the history of our time. It marked a remarkable moment in the history of mankind and an amazing achievement for one man – Neil Armstrong. Think about the incredible risk that he took to literally blaze a trail through space, onto the moon and into the history books. And think of all the preparation, effort and teamwork that went into the Apollo mission.

(The presenter should read out extracts from this famous MLK Jr speech – or play a section of the original speech)

"I have a dream that one day this nation will rise up and live
out the true meaning of its creed.
I have a dream that one day on the red hills of Georgia, the
sons of former slaves and the sons of former slave owners will
be able to sit down together at the table of brotherhood.
I have a dream that my four little children will one day live in
a nation where they will not be judged by the colour of their
skin but by the content of their character.
I have a dream today!"

These words were delivered to over 250,000 civil rights supporters from the Lincoln Memorial in Washington in March 1963. This is one of the most acclaimed speeches in American history. The words were spoken by Martin Luther King Jr – a Nobel Peace Prize winner in 1964. His work transformed the lives of so many by putting an end to racial segregation in the USA.

Although their achievements are very different, these two men have many things in common. It is not only the place and time in history that unites them but key personal qualities that set them apart from others. They were leaders of their time and like the other people whom we heard about at the start of the assembly, have certain key qualities that enabled them to leave such a large footprint on the paths of our history.

MAIN POINT 3: DEVELOP YOUR PERSONAL QUALITIES

So what qualities did these people have? First of all, they had a clear vision and a strong belief in what they were trying to achieve. They were determined and were prepared to try and try again if their first attempts failed. Many of them showed great bravery or made personal sacrifices for the good of others. Many of them will have faced opposition and criticism from people who did not share their vision. At times, this would have been very hard to take, but still they persevered. These are the qualities that made these people the pioneers and innovators of our time. All leaders in their own way - but what exactly does 'leadership' mean?

Let's reflect on that for a minute - what does it take to become someone who has the ability to lead others. Here is a quote to think about:

"A leader is one who knows the way, goes the way and shows the way." [John C Maxwell].

In other words, they have a vision; they follow their belief, and by doing so convince others to do the same.

The people we have talked about today all have a place in our history books but remember that you can lead at all levels in life and society without necessarily being famous.

You do not need to be the next president of the United States to be a good leader. It might be that you become captain for your team or House at school, or become a Prefect or have another student leadership role – or simply set a good example as a dedicated student or as a decent person.

SUMMARY

Today we have thought about some famous people in history and the impression they have made on our lives. We have considered the qualities these people have and what it means to be a leader. We have also had the opportunity to reflect a little on what positive things we can do that people will remember _us_ for.

CLOSING:

So what are your ambitions and dreams? Where will the future take *you*?

Someone in this room might be a future politician or author or famous footballer. One of us might invent something that nobody has thought of before or become a success in some other way. Which person in your class do you think will be the most successful? What is it about them that makes them special or different – that sets them apart just that little bit?

It is important to remember however that just being famous does not mean that we are a success – because people can be famous for very different reasons – some good, and some bad.

Leadership and success can also come quietly into our lives. This might be as a good friend, a great parent, an influential teacher or just someone who works hard and is successful at what they set out to do. What is important is that we make a difference – however small, in some positive and constructive way. So, please take a minute before you leave this assembly to ask yourself – how will I make my mark?

Thank you for listening.

5

DEMOCRACY

AIM: To encourage an awareness of democracy, this assembly discusses the history of voting in the UK and what it means to live in a democratic society.

Optional Props: legal rights age limit cards; photographs of political party leaders; party emblems.

Time: Approximately 15 minutes.

HOOK

Can you help me with some problems?

I want you to imagine that it is the weekend. It is Saturday morning and you have met up with a group of your friends. You are all trying to decide how you are going to spend the day. There is lots of discussion going on. Some of you want to go the cinema to see the latest film release, others in the group want to go bowling, some think swimming might be fun whilst a couple of you just want to have a wander round the shops. There are lots of choices – how do you decide what you are going to do?

Now let's take another situation. The school is thinking about having a shorter lunch time but allowing pupils to leave school thirty minutes earlier. How could we decide whether this is a good thing to do?

Finally, it is Saturday night and the nation is gripped by the anticipation of the latest X factor final – so how do we get our winner?

(The presenter could adapt these examples to any other situation – for example; 'Which team is going to win the Premiership?' etc..)

The answer to all of these questions is that we would take a vote and go with the majority – that would seem to be the fairest way to decide what to do. Voting in this way would take account of everybody's views and opinions. It is a good example of what it means to be democratic. So what does democracy mean and what is it like to live in a democratic world – let's find out more.

OVERVIEW

In today's assembly, I want us to think about what it means to live in a democratic society and what it means to have the right to vote. We will also consider the history of voting in the UK and how our country operates today as a democracy. Finally, I would like us to think about examples of how we can be democratic, and reflect for a moment on what it might be like to live in a world where democracy does not exist.

MAIN POINT ONE: THE HISTORY OF VOTING IN THE UK

Let's go back to our opening situation for a minute and imagine that the decision about how to spend your Saturday had been made in a very different way. How would you have felt if one of your friends had taken charge, made the decision and simply told everyone what they were going to do? You would probably feel that this person was acting unfairly and might feel that your views had not been taken into account. It would not have been a democratic way to make the decision. In life there are times when people have to take the lead and make decisions for everyone – this is what it can mean to be a leader at times. Sometimes they have to act quickly and make difficult decisions. However, the fairest thing to do in our Saturday activity situation is to take a vote, and go with the majority decision. That basically, is the idea of democracy.

We will all vote lots of times in our lives and for lots of different things. It might be to make a simple decision amongst friends or in class, to elect a representative for school council or even to vote for our favourite celebrity to stay in a reality TV show. What is certain is that once you are older, you will all have the right to vote in elections to decide who should represent you in governing our country. This is what it means to live in a democracy. It gives you the right to express your views and cast your vote in elections if you chose to do so.

Now let's think about how old you have to be to vote in this country. In the UK, do you know how old you have to be before you can get married (with parents' consent); learn to drive; work full time; or work part time? *(The presenter could do this with volunteers matching cards for the legal right and age limit. The answers are 16; 17; 16 and 13 respectively.)* So, how old do you think you have to be before you can vote? In this county, you will be able to vote once you reach the age of 18.

We could be forgiven for taking it for granted today that we have the right to vote once we are 18 but this has not always been the case. It may surprise you to know that when the General Election of 1708 took place only 3% of the population were eligible to vote. Hardly a majority and hardly democratic!

Fortunately, reforms followed although not always as quickly as hoped. By 1832 the situation was better, but still 6 out of 7 men could not vote and women had no voting rights at all. Doesn't this seem alien and almost unimaginable in today's world of equal rights? Can you guess how long it took until all men had the right to vote – what year do you think it was – any ideas?

It was not until after WW1 (1918 – some 86 years later) that there was universal suffrage in the UK – that is, that all people had the right to vote. However, this was *still* not strictly true because it took another 10 years until women under the age of 30 were also allowed to vote! The campaign of the women suffragettes is well known and is a topic for a different assembly, but what we do need to appreciate is that many people campaigned hard and sometimes made personal sacrifices to win our right to vote. Whether in years to come you exercise your right to vote will be a matter of personal choice but it still important to recognise that this right has not always automatically been there.

MAIN POINT 2 – DEMOCRACY IN THE UK TODAY

I want you to think a bit now about how democracy operates today in the UK. Can anyone tell me the name of the Prime Minister and what political party he represents? *(Field responses from the students.)*

That was pretty easy really, so let's try another. *(Field some questions at this point. You could extend this segment by showing photographs of the main party leaders or political party emblems to see what pupils recognise)*

Q: Who can tell me the name of the Leader of the Opposition?
Q: The name of that political party?
Q: What the opposition party does?
Q: Does anyone know the name of the 3rd main political party?

So, what have we learned?

We know that there are three main political parties in the UK – Conservative, Labour and Liberal Democrats. There are also smaller parties who may also have elected representatives in the

House of Commons such as the Green Party or UKIP. In recent times, it has usually been either the Conservative or the Labour party that has the majority vote and has formed a government.

There are 650 seats in the House of Commons – each representing one area of the UK. A political party must win 326 seats in order to have an overall majority – or just over 50%. However, in 2010 we had an unusual situation in that no party won an overall majority and there was what we call a 'hung parliament'; the first since 1974. As a result the Conservative party and the Liberal Democrat party joined forces forming a coalition government to give an overall majority over Labour which is why we had a Prime Minister and a deputy Prime Minister from different political parties (and often with very different political views!)

So, how are these parties elected and how does voting work in the UK? A General Election is when the country gets to decide who should form the new government.

(Optional question)
Q: Does anyone know how often General Elections are held? Any ideas?

A: Candidates campaign for weeks before the election – lamp posts are adorned with posters, leaflets drop through our doors and our TV screens are flooded with party political broadcasts. Promises are made, manifestos are read and politicians debate passionately on televised debates– all in the hope of winning our vote!

One party is currently thinking of lowering the voting age to 16 to increase the number of people who can vote – do you think this would be a good idea? Go back to those ages at the beginning of the assembly – you can get married at 16 and fight for your country in the army but you cannot vote. Some food for thought perhaps?

Then for one day only, everyone in the country has the chance to vote. Polling stations open early in the morning until late at night.

Votes are frantically counted to see which constituency will be the first to return their result. Candidates wait tensely and eagerly as the results gradually come in. Election night can be a time of exaltation and celebration as well as a time of disappointment and defeat.

Each seat is won by the local candidate with the most votes – the one who is literally first past the post. The results for each area are tallied. Then, when it is clear which party has won the largest number of seats or, once they reach the magic number of 326, the results are announced and a new government is formed. The new Prime Minister appoints his ministers to government departments and off they go - until the election process begins again five years later. That is how our voting system works.

MAIN POINT 3 – HOW DOES DEMOCRACY WORK FOR YOU?

There are ways in which you are involved in democracy in our everyday lives even though you do not yet have the right to vote. Let's think a minute about our school. We all vote at the start of each year for members of our form who we think will do a good job in representing our views on school council. They meet with other representatives on a regular basis and discuss issues that are important for our school. Your form representatives are able to take forward your views. You might decide now that this is something that you would like to become involved in or perhaps you could represent your form on one of the other student forums in school. It is certainly something worth thinking about to give you a chance to develop your skills and also get you involved in making decisions that will benefit yourself and your peers.

School Council is not only an example of student leadership but is also a forum through which you, through your elected representatives, have the opportunity to present your views as decisions are made. It also allows you to have feedback as to why certain decisions are made – an example of people having their say and acting in a democratic way.

(The presenter may wish to note some examples of democratic councils or groups working in the school environment and some of their successes.)

SUMMARY

So we have thought a little more today about the term 'democracy' and what it means to live in a democratic society. We have learned how the political system works and how, when we are older, we will be able to contribute to this. We have also had the opportunity to consider situations in our lives and in our school at the moment where we are able to be democratic and share our views and express our opinions.

CLOSING

There are still countries today that are not truly democratic. There may be leaders in power for long periods of time where elections do not take place and people do not have the opportunity to express their views. Imagine what it would be like to live in a country like this. We hear a lot of debate about the political decisions in our country at times – it is almost as popular as discussing the weather – and there may be times that we don't always think the right decision has been made. However, let's reflect and be thankful today for the fact that we do at least have the opportunity to express our opinions and make our views heard.

If we need to and choose to, we can at least, take a vote.

Thank you for listening.

6

MAKING TRADE FAIR

AIM: To encourage an awareness of fair trade, this assembly asks students to consider where their food comes from, why we need trade and what is being done to make trade fairer for people in LEDCs (less economically developed countries).

Optional Props: empty ingredient packets for chocolate cake; copy of FairTrade symbol..

Time: 10 minutes approximately

<u>HOOK</u>

Are any of your hungry? I hope not because I want you to imagine that I have baked a big chocolate cake. I have put jam and cream in the middle of my cake with some slices of banana. I have decorated it with chocolate flakes and sprinkles. I am looking forward to having a piece of my cake when I get home from work with a nice cup of tea or coffee.

Now what did I use to make my cake? Here is my shopping list:

I used butter, sugar, eggs, flour, cocoa powder, vanilla essence, chocolate, jam, bananas and sugar for my cake. I will use coffee, teabags and milk for my cup of tea or coffee. Delicious!! Feeling hungry now?

(The presenter could use empty packets or actual ingredients and have a volunteer take them out of a shopping bag as you go through your list.)

Q: Now can anyone tell me which of these ingredients come from countries outside the UK?

The ingredients which started life outside the UK are: cocoa; coffee; tea; sugar; vanilla and bananas. So let's find out what this has to do with our assembly today.

OVERVIEW

Today, I want us to think a little about where our food and other goods come from. Many products we use daily have travelled thousands of miles before they reach us. We rely more and more on trade with other countries – we buy goods *from* them and sell our products *to* them. This is called 'trading'. But how does trade really work and why are there concerns about trade being unfair for some countries? We are also going to consider what is being done by organisations in an attempt to make trade fairer for some groups of people around the world.

MAIN POINT ONE – UNDERSTANDING TRADE

(If you have brought ingredients in, you could ask for volunteers to put names of countries or flags on them as you go through your list).

I want to take you back to my ingredients for a minute so you can reflect on where each have each come from. I started off with cocoa which came from Ghana in Africa; then coffee and vanilla which originated in Brazil; my tea was grown in India; my sugar cane came from Mexico, and the bananas started life in the tropical Windward Islands.

Next time you go shopping, have a look at the labels on the foods that you are buying in the supermarket and think about all of those that come from outside the UK. What would it be like if we did not trade with other countries and could not buy these foods? What foods would you *not* be able to buy or eat that your regularly enjoy? How would the contents of your fridge, fruit bowl or cupboards be different? Imagine, for example, a world without chocolate or sugar or tropical fruit?

So, what does the word 'trade' mean? Think of times when you have traded something – like football cards or maybe sweets – it means simply that you swapped one thing for another. Trade between countries is the same really – quite simply it is the exchange of goods between countries – except that money also changes hands. We buy things from other countries called 'imports' and sell things to other countries called 'exports'. Imports = IN. Exports = OUT. In this way countries rely on each other and become interdependent. It means we can get the goods we need which we cannot produce ourselves. It also means that we can make money by selling the products *we* make which *other* countries need.

In the UK and other more economically-developed countries, the main *imports* tend to be primary products like the food stuffs described earlier. These are usually quite cheap for us to buy. The goods that developed countries *export* are usually manufactured goods like cars and electrical goods. These tend to be more expensive goods. In less economically-developed countries, it simply works the other way round –*they* export their primary goods and buy manufactured ones as imports. Have you got that? That is basically how trade works. We agree to sell our manufactured goods to them, and they agree to sell their primary products to us. The problem arises however when it comes to paying for these goods, because workers in developed countries tend to earn far more than workers in other poor countries – which in turn makes products either less-or-more expensive to make and to buy, depending on where you live.

(To model this, ask for a volunteer to come to the front of the assembly. Give them a coconut to sell. Explain to the assembly how hard that person has worked collecting coconuts – then offer them just 2p for their coconut and see what their reaction is. Now produce a picture of a car and tell them they need to give you £10,000 if they want to buy this car. Now ask everyone to reflect on how many coconuts they would need to sell and how long they would have to work to be able to buy the car? 500,000 coconuts is the answer – or enough time to fill the whole assembly hall with coconuts!)

MAIN POINT 2 – WHY IS TRADE UNFAIR?

More economically-developed countries like the UK make *more* money from their trade than less economically developed countries do. But why does this happen? Think back to what we already know about imports and exports. Developed countries sell expensive manufactured goods and buy cheaper primary products like food. Therefore, they earn *more* money from their exports than they need to spend on their imports. They have a good balance of trade. For poorer countries, the reverse is true. They do not have the infrastructure or facilities to make expensive consumer goods such as vehicles – so they have to import those from us at high costs, while selling their exports to us at a relatively low cost. This can be a hard cycle to break and as a result, many less economically-developed countries stay poor and find it harder to develop.

The effects of this trade cycle are felt by individuals too. Often the price that the farmer and workers in less economically-developed countries receive is very low. In addition to this, if the price for that product falls on the world market during a particular year, it means that they receive even less for their goods during that time. Both of these factors mean that workers and farmers in less economically developed countries do not always have sufficient money to meet their day-to-day needs. Is that fair? Let's look at an example to see what you think.

Consider a banana – it is selling for 30p. **Q: How much of this 30p goes to the banana worker? What do you think?**

A: You might be surprised to find that the amount is very low – just 1p of the final price gets paid to the banana worker in the producing country. The rest goes to the importer; the shop/supermarket; the shipper and the plantation owner. Is that fair? What do you think?

There are many people who do not think that this is a fair situation and are taking action to try to change this situation – let's see what they have done

MAIN POINT 3 – MAKING TRADE FAIR

Have any of you heard of 'FairTrade'? You may have noticed when you are shopping that some products carry a FairTrade mark . *(The presenter can hold up a copy of the symbol at this point if you wish)*

The FairTrade mark was first introduced in 1988 and the FairTrade Foundation was established in 1992. Having the FairTrade mark on a product is like a promise to give a better deal to the farmers that have produced it.

I want to explain to you how FairTrade works. The consumer (in rich countries like ours) might pay slightly more for the product but the good thing is that the farmer then receives a guaranteed *minimum* price for their goods. This means that they are better able to care for themselves and their families as well as make plans for their future.

FairTrade also means that a premium is paid that can be invested in community development projects such as roads, clean water, education and healthcare projects – all of which will improve the quality of life for the farmers and producers living in less economically developed countries.

FairTrade is also committed to sustainable development and

securing better working conditions for farmers – so all in all things look far brighter for farmers of FairTrade products.

You may be wondering what you and others can do to help. There are lots of ways in which people are currently getting involved in FairTrade. Some may decide to spend a little extra and use their power as a consumer to buy FairTrade products in the shops and supermarkets. Others may sign petitions; take part in events to raise awareness; make donations or carry out fundraising events. Every year there is a 'FairTrade Fortnight' which runs from 25th February to 10th March. It is even possible to become a FairTrade Town, a FairTrade Organisation or even a FairTrade School. Currently, there are over 800 FairTrade schools in the UK, all of which have given a commitment to support FairTrade. They have a FairTrade committee, and promote FairTrade products. They learn about FairTrade in lessons and organise events both in school and the wider community. So, as you can see, there are many ways in which people can become involved in promoting FairTrade.

SUMMARY

We have seen today how the world is interdependent and that we rely on other countries through trade. We need to buy things that we cannot produce ourselves and sell the things we make to earn money for our country. However, that trade works better for some countries than others. People are taking action to solve this though FairTrade.

CLOSING

Next time you are out shopping, think about all the products that you buy that come from other countries. Reflect on what products you would not have if we did not trade and were not interdependent. You can also have a look out for the FairTrade mark and see how many goods you can find. It is good to know that the farmers that produced those products are now getting a better deal as a result of FairTrade.

Maybe you will decide to find out a little more about the FairTrade Foundation and what people are doing to try to make trade fair. You might decide to get involved and take action in some way either individually, or collectively as a school. Either way, you now know more about this important issue and are aware of the choices people can make to make trade fair and improve the quality of life for poor farmers and producers in many other parts of the world.

As we leave today, please think about this motto used by the FairTrade Foundation:

"Act now for the future of food, together we can make food fair".

Thank you for listening, and don't forget about those coconuts!

7

SOCIAL MEDIA
FRIEND OR FOE?

*AIM: To raise awareness of the pro's and con's of internet use,
and especially the dangers associated with social media.*

Time 5 to 10 minutes

<u>HOOK</u>

**Q: Can you raise your hand if you have ever used social media?
Facebook, My Space, Twitter or any of the other ways of
communicating by using modern technology?**

I am sure for most of you the experience has been fun! A great way
of keeping in touch with friends sharing gossip, meeting new people
or just reading jokes.

Here are a few fun posts made on social media recently.

> "I was walking in the park the other day and I found a Justin
> Beaver ticket nailed to a tree so I took it, you never know
> when you will need a nail?"

> "When I die, I want to go peacefully like my grandfather
> did–in his sleep. Not yelling and screaming like the passengers
> in his car."

"Always borrow money from a pessimist. He won't expect it back."

Yes, this is all a bit of fun but there is a darker side to social media.

MAIN POINT 1: WHO ARE YOUR *REAL* FRIENDS?

Can you raise your hand if you have more than ten friends on your social media site? More than 50? More than 100? More than 500? How many of those friends do you really know well? By that I mean how many of those 'friends' have you actually met? Are you 'friends' enough to be able to call them up for a quick chat? Probably not – right? In many cases most of those 'friends' are probably strangers that you know very little about.

What could happen if some of those friends are actually criminals who would know when your house was empty when you tell all your facebook friends that you are off on holidays?

Others could use your name address and date of birth to steal your identity, perhaps enabling them to take money from your bank account.

Some of the people you are talking to may have created false identities, so never go and meet someone you don't know. Just think about the risks. There are cases where schoolchildren met up with so-called 'friends' they met on facebook. They thought they were both the same age and went to a school down the road. But when they met in person, the new 'friend' turned out to be an older person who had been lying about who they were.

Q: So, why would they do that? Why would they want to meet you alone?

A: There have been cases where children have been abducted, attacked, and even murdered in situations like this.

You also need to remember that whatever you post on social media can be around the world in minutes. Anything you upload should be suitable to be shared with total strangers, future employers, your teachers, parents etc etc. Because once it is out there you cannot get it back.

MAIN POINT 2: BE CAREFUL WHAT YOU SAY ONLINE

Let me tell you a story about Lily..

Lily enjoyed using her social media site when she was a student at school and chatted to her friends, and swapped gossip and pictures. She fell out with one of her friends and sent some nasty comments using this social media site. She encouraged her other friends not to speak to this person, asking them to ignore her at school.

After finishing school, Lily trained as a holiday rep and was looking forward to travelling the world. She applied for her ideal job working for a leading UK travel firm.

At the interview, she was asked the question, 'Why was Lily the right person for the job?' Lily replied that she, "..was kind, considerate, helpful, and liked working with people." She said that she "..always got on well with everyone."

This somewhat surprised the interviewer who had checked out Lily on her social media page. She pointed out the very unpleasant way she had treated one of her friends and felt that Lily would not be a suitable person for the job.

What does this story tell us about our use of the internet? Well, we learn that we have to be very careful about what we post online. It also tells us that what we put on the internet will stay there for a long time and could very well catch up with us in the future.

Think about your internet footprint? What message are you leaving about yourself to others?

Lily was usually a pleasant girl who did well at school and was popular with most of the other students. She did not consider herself to be a bully. **Q: Would you agree?** *(Ask assembly - optional)*

Although Lily was not bullying anyone at school, she could be considered a cyber bully. The message she had sent and encouraged others to send really upset her old friend who became anxious and then started to self-harm.

Cyber bullying is defined as sending text messages or e-mails that are designed to upset someone else. Believe it or not, it is now a crime for which you can be arrested.

(Optional question for selected students)
Q: What advice would you give to a friend who is a victim of cyber bullying?

So, the general advice out there is:
1. Tell an adult you trust if you are being cyber bullied
2. Don't respond or retaliate to bullying messages – it will probably make things worse. Cyber bullies thrive on getting a reaction out of you – so don't give them the satisfaction.
3. Block users who send you nasty messages
4. Save abusive emails or messages (or texts) you receive
5. Make a note of dates and times you receive bullying messages, as well as details you have of the user's ID and the URL.
6. Don't pass on any cyber bullying videos or messages – this is promoting cyber bullying
7. If you are bullied repeatedly, change your user ID or profile, and use a pseudonym or some other made-up name that doesn't give any information away about you
8. You can talk to someone at Child Line for free on 0800 1111.

You will probably be aware of 'Safer Internet Day' which was launched in February 2015 and involved over 100 countries and 800 organisations. The theme for the day was, "let's create a better internet together". This day encouraged us to think about how we can use the internet in a responsible way.

We all have the right to use the internet but with rights come responsibilities. Make sure that you take your responsibilities seriously; that you are a good online citizen and that you follow the appropriate advice to stay safe on the internet.

Let's see how much you already know with our quick quiz.

(The following questions can be fielded to the assembly)
Q: You receive an email offering you a prize, you have never heard of the company what should you do?

A1. Contact the company and get you £5000 prize sending your address and phone number and bank details as requested.
A2. Forward the email to a friend so they can get the money as well!
A3. Delete the email.

Q: You receive a friend request from Jimmy and Phil that pops up on your social network page. You are not sure who these people are although they say they are a friends of friends. What should you do?

A1. Accept them as friends
A2. Ignore the request
A3. Go and meet Jimmy and Phil in the park where they said they would be at dusk

Q: You have a photo of a friend being silly at a party what should you do before you upload it?

A1. Request permission from the friend to upload the photo.
A2. Send it to some mutual friends and ask what they think.
A3. Enhance the image quality and enlarge it.

CLOSING

The internet has revolutionised our lives. You can now communicate with anyone in the world almost instantly. It is a map, encyclopaedia, a phone, a film theatre, a shopping basket and much more. You can see what is happening all over the world via web cams -something that your parents would have thought of as Science fiction – when they were at school.

However, is the internet friend or foe? Who would think that turning on your computer to access the internet could be so dangerous? Many of these dangers are also often hidden. Hopefully this assembly made you aware of some of these risks but has also shown you that there are practical steps you can take to keep yourself safe.

Finally remember the internet is the only Human invention that can never really be turned off. What *goes* on there - *stays* on there!

Thank you for listening.

8

READ ALL ABOUT IT!

AIM: This assembly asks pupils to reflect on their favourite books and shares some interesting facts about books. In doing so, it reinforces the importance of reading and hopefully, will encourage students who do not read to go away and pick up a book today.

Time: 10 minutes approximately

HOOK:

There is an old saying that says, 'There is a book in everyone". Raise your hand if you have heard this saying before. Have you ever stopped to wonder what it might mean?

Well, it means that if we put our minds to it, we could all write a book. Can you imagine yourself writing a book? What sort of book would you write? Would it be a fiction book? A romance, a thriller, a horror, a murder mystery or a fairy tale maybe? Perhaps, it would be a non-fiction book. Perhaps you would write about one of your interests or hobbies?

How many people in this assembly might one day write a book I wonder? Who knows, we may have the next Tom Gates, David Walliams or J K Rowling sat right here in this room!!

<u>OVERVIEW</u>

So, in today's assembly, I am going to ask you to think a little bit more about the part that books play in our lives. We will consider our own favourite book and what makes it special. We will also think about reading, what our world would be like if we could not read and how reading brings many benefits to our lives. Finally, we will reflect on the experiences of children in different parts of the world who do not have the chance to go to school, who cannot read and for whom the magic of books is something they have yet to discover.

<u>MAIN POINT 1: READING IS A PRIVILEGE.</u>

So let's start with a quick quiz about books:

Q: Who can tell me the most published book in the world?
A: The Bible

Q: Which author has sold the most books?
A: Agatha Christie – 4 billion

Q: The world's largest library is in the US. It is called the Library of Congress. Do you know how many miles of bookshelves they have?
A: 800 miles – that is nearly as far as from Land's End to John O'Groats!

That is a lot of bookshelves!! Apparently, that library contains over 150 million items! That's a lot of books! How long would it take to read all of those items do you think?

Now, can you remember learning to read? Can you think back to being four or five years old and the time that you managed to read your very first book? It was probably only a few words long but I am sure you felt very excited and proud. For the first time ever, you were able to decipher the meaning of the letters on the page. What you won't have realised at the time was that by learning to read you were unlocking a door to a very special place. A special place that you could revisit at any time and age for many years to come.

Now not all of you will enjoy reading. Many of you might prefer to watch TV, play sport or play on your games console and that, of course, is fine. However, what is important to remember is that being able to read is a gift to be cherished and something to be celebrated.

MAIN POINT 2: BOOKS ARE INTERESTING.

In order to celebrate reading, we hold World Book Day each year. In 2015, World Book Day was held on 5th March with 100 countries taking part. How did this school take part? Many teachers and children in our country celebrate by dressing up as their favourite literary character and taking their favourite book into school. You might have met an author or poet, taken part in quizzes and games or received a book token to enable you to buy your own World Book Day book.

Q: I wonder what your favourite book is?
(Take some suggestions from the audience at this point)

Q: Who is your favourite character in that book? I wonder why you like this book so much? What made you turn the pages?
(You might want to involve teachers as well as students at this point)

Q: How would you recommend this book to a friend?
(You may wish to use this example or include your own favourite book)

My favourite book when I was growing up was *Charlie and the Chocolate Factory* by Roald Dahl. I could not wait to get home from school to pick the book up again and find out what had happened next. I loved Willy Wonka and his amazing chocolate creations. They all sounded so tantalising. I found it very funny when Augustus Gloop fell into the chocolate river and got sucked up the pipe. I was enchanted by the umpa lumpas and their silly songs. I felt sorry for Charlie Bucket with his meagre diet of cabbage soup. I adored the moment when he opened his chocolate bar to win his golden ticket – it was just like it was happening to me. What a wonderful magical world this was; that I could escape to every day when I got home from school.

Reading offers us an opportunity to escape to different places and to use our imagination to identify with the characters in our books. However, reading also benefits us in many other ways too. Here are some quotes worth thinking about:

> *"The more things you read, the more things you know. The more things that you learn, the more places you'll go."*
> Dr Seuss

> *"Reading is to the mind what exercise is to the body."*
> Joseph Addison

> *"Reading gives us a place to go when we have to stay where we are."*
> Mason Cooley

> *"Today a reader, tomorrow a leader."*
> Margaret Fuller

> *"A house without books is like a room without windows."*
> Horace Mann

So what can we learn from these quotes. We learn that reading helps us to gain knowledge and understanding. It can make us more informed. It offers a way to switch off and relax. It will teach you new vocabulary and help you to spell new words as well as develop your skills of both concentration and communication. Imagine a world in which you were not able to read and how difficult that would be. We read all the time in our day-to-day lives – whether this is finding a destination on a bus, or following a recipe for our favourite meal, or choosing a holiday from a brochure, or reading a road sign to find our way – where would we be in the world if we could not read?

MAIN POINT 3: READING IS A GIFT.

We are so lucky that we have the opportunity to go to school and learn to read. For some of us learning can be tricky but we are supported and helped by our teachers and parents. We have libraries of books we can borrow both in our towns and in our schools, and iPads, computers and kindles at home that offer us even more choice. However, this is sadly not the case for some children in other countries. Let me tell you about a young boy who lives in Africa.

Aadan is 11 years old. He lives in a village in Somalia, Africa. His family is poor and cannot afford for him to go to school. He did attend school in the local village but had to leave when he was 8 years old to look after his younger brothers and sisters. Now, he helps his father on the farm and spends 3 hours each day just looking for water and collecting firewood. His life is hard and he has little time for play.

Aadan has a book that his grandma gave him before she died from dysentery. He loves to look at this book in the evenings when he has finished his work for the day. It is called 'Treasure Island'. He likes to look at the pictures that are contained in the book, but he can't read the text. He longs for the day when he will be able to read the book for himself. Aadan cannot write either. Neither can 60% of the people that live in his country.

Can you imagine what it might be like to live in a country where you do not have the opportunity to learn to read? Can you reflect for a minute how lucky you are that you are able to read and enjoy the many benefits that it brings. Imagine if this was *your* way of life – I know which one I would prefer.

SUMMARY

Today we have had the opportunity to consider the value of literacy and the privilege of having free access to books. We have looked at what makes reading special for you and have reflected on how fortunate we are in our country to have the opportunity to learn to read and write. So let's make this opportunity count each and every day.

CLOSING

We have various strategies in school to encourage young people to develop their reading skills because we recognise just how important this is – not just to enrich our lives but also to prepare us for doing well in our examinations and later in the world of work. So, if reading is something that you don't do much of right now, please stop and reflect. Let's think about those children who do not have the chance to read and who do not have the gift of books. Let's remember why it is important to read and the many benefits that reading brings. If you are still not sure, have a think about this final quote from Rachel Anders – "the journey of a lifetime starts with the turning of a page".

Happy reading, and thank you for listening.

9

THE IMPORTANCE OF LAUGHTER

AIM: To explain why laughter is a good thing but how humour can also have a darker side.

Time 5 to 10 minutes

<u>HOOK</u>

I would like you to think about the funniest jokes you know. I rather like this one. *(Presenter should use any other joke as appropriate, as a hook)*

Wolverhampton *(insert name of local underdog soccer team)* was playing Aston Villa *(insert name of local major team)*. A visitor to Wolverhampton went with his friend to watch the match. After about twenty minutes Wolverhampton got a corner. The crowd went wild, screaming and cheering, standing up and stomping their feet. "Wow," said the visitor to his friend. "That's amazing. What happens when they score a goal?" His friend said, "I have no idea. I've only been coming here for ten years."

MAIN POINT 1: HUMOUR IS COMMON TO US ALL.

One of the easiest ways to break the ice when you meet strangers is to share a joke – provided it is funny and it makes them laugh of course! The British are famous for their sense of humour, which comes in many forms; from satire to sarcasm, and from slapstick to stand-up.

Some of our TV comedy series are famous throughout the world. Think of *Monty Python, Blackadder, Only Fools and Horses* and *Mrs Brown's Boys*. We even like our quiz shows to be funny, and satire is especially important to us. Where would we be without *Have I Got News For You* and *Mock the Week*?

MAIN POINT 2: HUMOUR IS SCIENTIFICALLY GOOD FOR YOU.

Lacking a sense of humour might not just be bad for your social life; it might also be harming your cardiovascular health. A new study shows that laughter actually increases blood flow in the body, proving the old adage that 'laughter is the best medicine' – at least when it comes to the heart.

We change physiologically when we laugh. We stretch muscles throughout our face and body, our pulse and blood pressure go up, and we breathe faster, sending more oxygen to our tissues.

People who believe in the benefits of laughter say it can be like a mild workout – and may offer some of the same advantages as a workout. *"The effects of laughter and exercise are very similar,"* says psychologist Steve Wilson. *"Combining laughter and movement, like waving your arms, is a great way to boost your heart rate."*

One pioneer in laughter research, William Fry, claimed it took ten minutes on a rowing machine for his heart rate to reach the level it would after just one minute of hearty laughter.

And laughter appears to burn calories, too. Maciej Buchowski, a researcher from Vanderbilt University, conducted a small study in which he measured the amount of calories expended in laughing. It turned out that 10-15 minutes of laughter burned 50 calories.

MAIN POINT 3: HUMOUR CAN HELP DEFUSE A DIFFICULT SITUATION.

For example; a man was driving up to a roundabout and ran into the back of the car in front. The guy whose car got hit was understandably very angry and aggressive. The driver who caused the accident calmly got out of his car and said, *"You know I really didn't intend to do that, I think my car is possessed. Have you seen my number plate? It is RBY 666. I'm sorry. Obviously the accident was my fault and I will pay for any damage."*

The aggressive guy found it hard to remain angry while laughing at the other guy's good humour.

A large part of our identity revolves around satire and parody – a very British form of humour that has been successful in challenging the establishment.

Q: How many of you will have seen or heard of the *Monty Python* film 'The Life of Brian'? Well, at its most basic, it's just a funny story about a man called Brian whose life didn't turn out quite how he expected. The film was extremely controversial however because it's also a parody of the life of Jesus. Many local authorities would not allow the film to be shown even though it had been certified by the British Film Board. Hundreds of religious protesters picketed cinemas in order to stop people from seeing the film.

At one point in the film, the revolutionaries, known as 'The People's Front of Judea' are complaining about the Roman occupation. Their leader, Reg, is stirring up support for the cause and asks his merry band the provocative question; *"What have the*

Romans ever done for us - eh!?" to which they unexpectedly reply: *"Well, they built roads, sanitation, brought us law and order, and peace."* Reg becomes agitated with these responses and says, *"Apart from these things, what <u>else</u> have the Romans done for us?"*

The list of good things the Romans have done for the locals then continues – *"..they brought us medicine, education, public order, and wine.."* Whereupon an exasperated Reg decides to change the subject.

This type of humour is called 'a parody' – a light-hearted take on a serious subject. But can you think of any other situations where people complain about things with little justification and without realising that whatever they are complaining about has actually benefitted them greatly?

For example, what have immigrants ever done for us - eh? Well, the fact is they have improved our food, contributed taxes and generally enriched our culture. Without immigrants for example, there would not be enough doctors and nurses to keep the NHS running.

We use satire to understand, examine and undermine political ideas. If you can see the silly side of something, it won't seem nearly as threatening and then it's easier to think clearly about what it really means.

Many famous politicians have been successfully machine-gunned with humour by Ian Hislop on *Have I Got News For You* and Ian's humorous approach can be just as effective as TV journalist Andrew Neil's hard-hitting strategy at eliciting the truth from difficult politicians.

MAIN POINT 4: HUMOUR CAN EVEN CHANGE NATIONAL POLICIES.

Comic Relief is one of the most successful charities in the world. By using humour in all sorts of ways, from getting people to sit in tubs of beans, to convincing celebrities to perform for free, people are then relaxed enough to then face the reality behind the real issues *Comic Relief* is trying to raise. This use of humour has produced hundreds of millions of pounds over the years and this money has been used both in the UK and in developing countries to make a real difference to real kids' lives. What a great use of humour.

Humour can bring people together. Shared enjoyment of shows like Mrs Browns boys enables people to share what made them laugh which produces a real feeling of belonging to the same group. Christmas Pantomimes are funny for adults and children but often there is an underlying message in all the silliness.

When people are frightened, someone making a joke can really take the tension out of the situation.

A parachutist going on his first jump from an aeroplane was very nervous and asked why am I doing this? The instructor said yes I always wonder why we are so stupid as to jump out of a perfectly serviceable aeroplane, although when I reach the ground I know why, and I want to do it again.

MAIN POINT 5: PEOPLE OFTEN USE BANTER TO EXPRESS AFFECTION.

"I am glad you are around Lofty" (Lofty was very tall) *"You are always the first person to know when it starts raining so the rest of us can get under cover."*

But what happens when this banter goes too far?

There is a dark side to humour too. When the joke changes from laughing *with* someone to laughing *at* them, then it's no longer funny. Cutting remarks which are intended to hurt someone's feelings are often disguised as 'jokes'. But there's a fine line between genuine humour and giving offence – and that line depends upon the reaction of the person who is the subject of the joke.

Sarcasm and personal insults are rarely funny. A bit of banter between friends that starts innocently enough can easily descend into offence or abuse if things become too personal.

SUMMARY

So yes, humour is fun and important to us in many ways. But when the humour becomes a method of demeaning or bullying someone else it is not a good thing.

CLOSING

So please think about what you are saying when you are making a joke. Because it's only worth telling if *everyone* can laugh at it.

Thank you for listening.

10

 **AMSTRAD**

SUCCESSFUL PEOPLE

AIM: To look at the lives of some successful people and to reflect on what made them successful.

Time 10 to 15 minutes.

HOOK

Would you like to be a multi-millionaire? Can you see yourself living in a huge mansion driving an expensive car or even owning your own island? Have you ever wondered why some people are very successful yet others have very little? What is it that makes people successful?

OVERVIEW

I would like to talk to you today about people who have been very successful, not by inheriting money but by their own hard work. People who started with very little, yet today they are household names.

Have you heard of *Amstrad*? The letters stand for Alan Martin Sugar trading, or you probably know him as Lord Sugar from the apprentice?

If you have travelled on *Virgin* trains or aeroplanes, you probably know who Richard Branson is.

If you have been to *The Body Shop* then you will have bought beauty products created by Anita Roddick.

Today we are going to look at the lives of these three people and see if there is something they have in common that makes them successful. If it was something you could bottle and drink I am sure we would all like to have some of *that!*

MAIN POINT 1: SIR ALAN SUGAR

Sir Alan Sugar was born on the 24th of March 1947.

As a child he lived in a council house in Hackney East London. His father was a tailor working in the East end clothing industry.

Alan attended Brooke house secondary school (now a sixth form college) in Upper Clapton. He worked at a local green grocer to make some pocket money.

He spent his savings (£50) on an old van and started selling electrical goods from the back of it. At the age of 21 he started Amstrad which he used to import and export electrical goods. From this he moved on to production, he used new technology to undercut competitors and was soon producing turntables and amplifiers at very competitive prices.

In the 1980s he moved into the computer market producing the Amstrad CPC 464 computer and word processor. He later produced low cost PCs. He then bought in to Betacom and Viglen computer service companies. He sold his interests in Amstrad to Bsky B in 2007. He is now estimated to be worth about 700 million pounds.

He is a very generous philanthropist giving money to charities such as Jewish care and Great Ormond Street hospital. He also donated 200 000 pounds to the British Labour party in 2001.

I am sure you have all seen him on the Apprentice and that series certainly gives an indication of his drive and determination to succeed.

MAIN POINT 2: DAME ANITA RODDICK

I am sure most of you know *The Body Shop*. This was founded by Dame Anita Roddick in 1976. This was a cosmetics company producing and retailing beauty products. She opened the body shop with the aim of making an income for herself and her two children.

It was one of the first companies to prohibit the use of ingredients tested on animals and one of the first to promote fair trade with third world companies. We will tell you about fair trade in a different assembly.

She opened her shop with the idea of providing quality skin care products in refillable containers all marketed in an honest way without the hype often associated with beauty products. She opened her second shop six months later.

Her shops caught the public imagination being natural products with no animal testing involved. Everything that could be recycled was recycled. Originally the bottles were recycled because they did not have enough to go round. But slowly recycling became part of the culture of her business

By 2004, the Body Shop had 1980 stores, serving over 77 million customers throughout the world. It was voted the second most trusted brand in the United Kingdom, and 28th top brand in the world.

In 2006 *L'Oreal* bought *The Body Shop* for 652 million pounds. In some ways Anita was unhappy about this as *L'Oreal* used animal testing. She justified her decision to sell by saying she would have some influence on the decisions made and may help to change the ethics of this huge organisation.

After a visit to Romanian orphanages she founded *Children on the Edge.* This charitable foundation helped children who lived in terrible conditions in the orphanages. It also helped children involved in conflicts, natural disasters and those affected by AIDS.

She wrote a book called *'Take it personally'* which encouraged fair trade and an end to the exploitation of workers and children in underdeveloped countries. In 2005 she decided to give away her fortune, about 50 million pounds.

She died of hepatitis C in September 2007. This is a disease that destroys the liver and she thought she may have contacted it from a blood transfusion after the birth of one of her children.

After her death her husband Gordon Roddick, founded '38 Degrees' in her memory, explaining, *"I knew what would make Anita really laugh would be to cause a lot of trouble."* Thirty Eight degrees is a pressure group that campaigns on the internet to change the law. They attempt to get enough people to sign a petition so that parliament has to debate it.

Anita had not created a new idea she just marketed her products in a sustainable way, supported fair trade and avoided animal testing. She saw a gap in the market and created a very successful business.

MAIN POINT 3: RICHARD BRANSON

Have you ever travelled on a *Virgin* train or aircraft? Bought a record from *Virgin* records or V2? Would you like to go into space as a tourist? If the answer is yes to any of these questions then you may be interested to know a little more about Richard Branson

Do any of you here not like school or find lessons difficult? Well you are not alone, Richard Branson struggled at school and dropped out at the age of 16. For his first business venture he started a youth magazine called *Student* it was run by students for students. He managed to sell £6,000 worth of advertising in his first edition. I wonder if any of you in this hall today would like to set up your own magazine?

He ran the magazine from a church crypt and from the same location he also started a mail order record business. This performed well enough for him to be able to open a record shop on Oxford Street in London. The record shop was so successful that he was able to open a recording studio in Oxfordshire. His first artist was Mike Oldfield whose album *Tubular Bells* was an instant success it was even used as the theme tune for the *Exorcist* film.

He went on to sign up *The Sex Pistols*, *Culture Club*, *The Rolling Stones*, *Genesis* and other aspiring groups of the time. This made *Virgin Records* one of the most successful recording studios of the 1980s.

He expanded his business with *Virgin Radio* and his second record company 'V2' which includes artists such as Tom Jones and Powder Finger.

The *Virgin* group now consists of more than 200 companies in over 30 countries. I am sure most of you know *Virgin Trains* and *Virgin Broadband.*

His latest venture is *Virgin Galactic* a space tourism venture. He is developing a sub-orbital space ship that will take passengers to the edge of space.
Again, Richard Branson had a small start to his business. He spotted a gap in the market for a student magazine and then went on to be one the most successful business people in the world.

Richard Branson has pledged to commit $3 billion, all profits from his travel firms over the next ten years, to the reduction of global warming.

He has previously donated to educational charities in Africa.

Branson started his first charity, *'Student Valley Centre'*, when he was only 17.

Richard is the patron of the *International Rescue Corps.*

In 2007, Branson formed *The Elders* – a small, dedicated group of leaders, working objectively and without any vested personal interest to solve difficult global conflicts – along with Peter Gabriel, Nelson Mandela, Desmond Tutu, Kofi Annan and Jimmy Carter.

When Saddam Hussein invaded Kuwait, Richard Branson organized to fly 40,000 blankets to refugees fleeing into Jordan. He single-handedly had to fight with Government ministers to make sure the supplies got to those who needed them.

SUMMARY

Well we have looked at three very successful people today. They all had a vision and determination to succeed. Perhaps they were lucky just getting the right product at the right time. But to be honest luck probably does not come in to the equation very much. If you want to succeed you have to be determined and willing to work incredibly hard. All have been very generous to charities. I wonder if you know of any very successful people who do *not* give generously once they have made their fortunes?

CLOSING

Well would you like to be one of these successful people?
If so, what qualities do you need to run your own successful business?

Could you start a business today is there was a product or service you could create and then sell on ebay?

Here are some ideas other students have come up with. Some of these ideas have been very successful.

- Could you build a web site for people or repair computers?
- Sell a soft toy you have made, or a picture you have drawn?
- Could you grow some plants from seed and sell them at the local garden centre?
- Grow some vegetables in your own garden?
- Produce a home-made solar heater?

The question is what sort of things do people really want to buy?

(The next section is optional. Assemblies should only be opened up for discussion if you are well prepared)

Have a chat with your neighbour for a few minutes and see if you can come up with any ideas.

(The presenter may then ask people to tell the assembly about their ideas - then, after discussion and feedback...)

Well thank you for all your ideas I wonder how many of you will be millionaires in a few years' time?

Well, if you do make lots of money don't forget your teachers on your Christmas list.

Thank you for listening

PAY IT FORWARDS

AIM: To explain the idea of 'paying it forwards'.
Time 5 to 10 minutes

HOOK

Have you ever done a favour for somebody when you have expected nothing in return? Has anyone ever done a favour for you for no reason except to help you out?

OVERVIEW

I would like to tell you a story. This is a true story about something that happened to a young woman who was out for a night on the town in London. Walking home on her own, Sandrine was attacked by a man on a bike. He punched her in the face, stole her bag and then pushed her. She fell backwards into the street and cracked the back of her head on the road. Sandrine could easily have passed for a drunk in the gutter but luckily for her, a man called Mark came by within a few seconds and realised she needed help. He rang for an ambulance and stayed with her, reassuring and comforting her until the ambulance crew arrived.

Afterwards, he rang the hospital to find out if she was ok and whether anyone was with her. It turned out that the woman had a very serious brain injury and the medical staff were not even sure whether she would live. Because the man acted so quickly, she made a full recovery. When interviewed some months later, she said that although someone had done a bad thing to her, the more important thing was that someone else had done something amazing for her. Although the woman was able to thank the man, there wasn't really anything else she could do for him. Instead, she decided to remember what he'd done, how it had made her feel, and to 'pay it forward'.

MAIN POINT ONE: GIVING IS GETTING

'Paying someone back' (as we all know) is usually when we return a favour to someone who has helped us in some way. But 'paying it forwards' means doing a good deed for someone else - even strangers - not because we owe them personally, but just because we can!

For example, a few weeks later Sandrine was on the London underground and saw a man stagger and fall at the bottom of an escalator. The man looked like he was homeless and maybe drunk, on drugs, or maybe mentally unstable. However instead of walking past she stopped to help him. A few weeks after that, she spotted a broken bottle in the middle of a suburban street. When she saw a car swerve to avoid it, she picked the bottle up and carried it until she finally found a rubbish bin. She didn't ask for any thanks for doing these things; she just did them because she could.

A few months later, she was once again rushing through London on her way to a job interview. It was her dream job and she was determined not to be late. As she was heading through Paddington station to catch her train, she noticed a child standing alone and looking fearful. She was desperate not to miss her train but something about the child didn't look quite right and she paused.

Looking more closely, she could see the child was trying not to cry. She couldn't see any adults in the vicinity who appeared to be with the child. She noticed a man had also stopped because he could see something was wrong but appeared to be hesitant to approach the child in case this was misunderstood. Sandrine's train's departure was being announced and she knew if she missed this train, she would be late for her interview. Although she was tempted for a moment to let the man deal with the child, she remembered what it was like to be alone with no-one to help and so went over to the child to see how she could help.

MAIN POINT TWO: GOOD ACTIONS HAVE GOOD CONSEQUENCES

So, it ended up being a really good thing that Mark stopped on the evening Sandrine was attacked. By saving her life and making her focus on the good, rather than the bad thing that happened to her, Sandrine passed his favour forward.

By stopping to help the man who fell at the bottom of the escalator, the ambulance crew and hospital medical staff were able not only to sort out his immediate injuries but they also discovered that he had been missing for several months. He suffered from a mental health condition and by Sandrine intervening when she did, he was able to get the treatment he needed and was reunited with his family and eventually able to return to work.

He later confessed that on the day Sandrine stopped to help him, he was suicidal and had made serious plans to end his life. Because she cared enough to help him, even though he was in a dreadful state, he felt his life must be worth something and decided to accept the help he was offered to get his life back on track.

Although Sandrine never knew this, if she had not removed the broken bottle from the road, the next vehicle to come around the corner was a mini bus full of children on their way home from a day at Thorpe Park. The bottle would almost certainly have punctured one of the tyres and possibly sent the mini bus skidding across the road into an oncoming lorry.

And finally, the child that Sandrine stopped to help was from Switzerland and although she spoke French and German, did not speak any English. She had become separated from her parents in the strange, busy station and didn't know how to ask for help. Sandrine was able to reassure the child and translate what she said to the station staff so that her parents could be found. The look of relief on both the child's and the parents' faces gave Sandrine the courage to ring the potential employer and explain why she would be late.

Now imagine what would happen if each of the three people Sandrine helped also did something nice for three other people (that's nine people). And what if those nine people also did something nice for three more people (that's another 27 people). Already, that's 40 people who have had something nice happen to them that they didn't ask for and where no return favour is expected. How would that change the world we live in?

MAIN POINT THREE: HOW ABOUT US?

Let's make it a little more realistic as we can't all go around doing the work of superheroes every day. But let's imagine you walk down the corridor and someone opens the door for you. Not for any particular reason - other than just because they can.

A little while later, you drop your pen on the floor and as your teacher is walking past, she quietly picks it up and puts it back on the desk for you. Again, she's not expecting anything in return.

At lunchtime, they've just run out of your **f**avourite food when the dinner supervisor tells you, *"Hang on a sec, I'll just go and get another tray for you."*

Later on, you're running for the bus, the driver sees you and waits.

When you get home and your little sister asks how your day was, you're in such a good mood that instead of telling her to leave you alone like you usually do, you find yourself smiling and saying, *"It was pretty good. How was yours?"*

SUMMARY

Well, there's a school of thought that says one of the best ways to improve your life and your world is to 'pay it forward'. It's not about making grand gestures all the time but about noticing what's going on around you and sometimes doing something nice for someone else *just because you can*. The great feeling you get from doing something for nothing is powerful and rather than taking my word for it, I would like you to test it out.

Over the next week, try to do five nice things for which you expect nothing back and see what happens because eventually, what goes around comes around.

CLOSING

P.S. About Sandrine's job – imagine her surprise when she walked into her interview, nearly half an hour late, and the little girl's mother was sitting across the table ready to interview her.

Yep, she walked it!

12

THE LEISURE SOCIETY

AIM: To explain the idea of a leisure society and discuss if such a thing really exists.

Time approximately 5 minutes

HOOK

With all the technology surrounding us today, isn't it wonderful that we save so much time with washing machines, computers, mobile phones, iPads and cars? I could go on but the list would be endless.

OVERVIEW

Imagine what it was like in the Victorian era when you had to wash everything by hand using a mangle to dry things. You would wear the same shirt for a week and only change the collar.

If you wanted a bath, you would have to fill it up kettle by kettle. If your homework included the question what is the square root of 473.678, you would have to look it up in a mathematical table then go through several complex calculations using only a pencil and your brain and even then it might not be right.

If you wished to communicate with someone who lived in another town, you would have to write a letter, address it, put it in an envelope, post it and pay for a stamp. If the person lived in another country, well good luck to you.

If you wanted to listen to some music you would need an orchestra. Holidays were only for the super rich. Even then, they only went to UK Holiday resorts.

Finding your way to places required things called maps. You had to learn how to read them and then figure out where you were on each section of your journey. When it came to cooking, if you wanted to have your lunch around noon you had to start cooking it hours in advance. If you wanted fresh fruit, you could only get what was in season.

Heating your house you would use a coal fire that you had to light in the morning keep fed all day and then clean out at night. Get the coal man to deliver your coal every week and the air was so polluted you could hardly breathe.

Toilets! We don't really want to talk about toilets but they were a necessary part of Victorian times too. In those days they consisted of a pit in the ground - outdoors - that you sort of squatted over. Very cold in winter and very smelly in summer, and we won't mention the flies and spiders. Of course at night you would have a chamber pot under your bed which you could pee or do worse in, marvellous!

Working hours were 12 hours a day with only a half day off during the week and half a day off on Saturday. Sunday you would spend all day in church. No days to sleep in all morning.

School was strict with children being beaten with large sticks if they didn't work hard enough; writing on slates for hour after hour; and reciting your mathematical tables for two months solid. Wearing a dunce's cap if you made a mistake - for total humiliation.

MAIN POINT 1: MODERN LIFE IS EASIER

Today of course things are much better, we live lives of luxury and leisure. Just because your mobile phone continues to receive emails 24 hours a day and your boss can text you 24 hours a day is no reason to think we are hard done by.

Why should your teacher not be able to facebook you on Sunday afternoon and tell you to get on with your homework? When computers first came out it was said that we would eventually have a 'leisure society' where computers and machines did all the necessary work.

Imagine you work in an accounts office. To add up the weekly totals would take 10 people 5 days of solid work. Introduce a computer which can do the whole thing in 2 hours, then you could give the people who work there 2 days off a week and pay them the same.

Computers were going to give us all lots more free time - a society where no one had to work. Yippee!

MAIN POINT 2: IS MODERN LIFE BETTER?

Why is it that we now work the longest hours of any European country and have the shortest holidays?

We also have more workers than most other countries. That includes people who work from home - which all sounds very nice except that you never get a chance to really switch off.

Surely modern washing machines save people lots of time? But is this really true? We now wash things every day instead of once a week (if you were lucky) in Victorian times. So today, even though we have machines to help us, we probably spend just as much time doing washing (and other jobs) as they did in Victorian times. It may not be as labour intensive to get the same results, but because of modern standards, time wise, it is very similar.

SUMMARY

So are we really better off with all of the technological advances of the last two centuries? Do we really have more leisure time, and are we any happier than several hundred years ago? Life is most definitely more comfortable but have we got the balance right yet?

CLOSING

As technology continues to advance at an ever increasing pace, it may be worth stopping and taking some time out to consider whether you are actually saving time with your new devices, or, whether those devices - as marvellous as they undoubtedly are - are taking valuable time away from how you actually want to live your life.

13

WATER, WATER, EVERYWHERE?

AIM: To encourage an awareness of the importance of water, this assembly explains how we use water, what problems water shortages can cause and gives practical tips on how you, the students, can help the situation at home and in school.

Optional Props: a litre bottle of water

Time: 10-15 minutes approximately

HOOK

Think about your day today since the time you got out of bed this morning. How much water do you think you have already used? No doubt you will have had a wash, flushed the loo, cleaned your teeth and had a drink - all of which will have used water. However, you will also have used water in many other ways, perhaps without even realising it.

Let me give you a couple of examples - water is being used right now in helping to make the electricity that is giving power to the lights in this assembly hall. Water has also been used to produce the paper that I have printed my assembly on. Come to think of it, water has been used to make the ink in the printer too! We use water constantly and in so many different ways, but do we realise just how much water we use and just how important it is not to waste it?

OVERVIEW

Today I am going to ask you to take a few moments to think about just how important water is and how we use it in our daily lives. I also want us to consider people in other parts of the world who are not as fortunate as ourselves and who face serious water shortages on a daily basis. We are also going to look at what you and your family can start doing today to help to conserve this vital resource.

MAIN POINT 1 – THE IMPORTANCE OF WATER

Q: Do you know how much of your body is made up of water as a percentage?

A: We are mostly water – nearly two thirds of our body is made up of water. We need water to drink when we are thirsty, to make our bodies work and to keep us fit and healthy. All living things, including us, cannot survive without water!

We also need water in our homes. We need it for cooking, cleaning, bathing, washing clothes and flushing the toilet to name just a few. I want you to think for a minute about how different life would be at home without a ready supply of clean water – what things would you *not* be able to do or have? Try to imagine for a minute a scorching hot day and you are really, really thirsty but you cannot have a drink because there is no water in the tap. What would that be like?

I have a question for you now.

Q: How many litres of water do you think the average person in the UK uses each day?

(The presenter should hold up a litre bottle of water so students can visualise the amount and to facilitate answers. This can be organised as a bit of a guessing game – give a figure and then ask who thinks it is higher, who thinks it is lower and so on. You could even involve your colleagues in the questions)

A: You may be surprised to know that we use on average about 150 litres of water per day (*imagine 150 of my bottles lined up on stage*) that's enough to fill two bath tubs - and most of it gets flushed away!

Every time we take a bath, a shower, flush the toilet, have a drink, do some washing; wash up; wash the car, water the garden - the list goes on and on - we are using water. Add up 150 litres for the people living in your house, in your street, in your school, in your town… and you'll soon see how much clean water we use every day.

Then think about the number of days in a year – 365 – and do your sums. That's a total of 54,750 litres of water we need every year.

We also use water to make things. These are the uses of water that are less obvious and might not spring to mind so easily. We use water to generate power; to cool engines in our transport systems and in industry –it takes thousands of gallons to make every car, for example!

So hopefully now we understand just how important water is to us in our daily lives and why we need to take extra care *not* to waste it.

MAIN POINT 2 – WATER SUPPLIES AND SHORTAGES

Approximately 70% of the surface of the earth is covered in water. Think about a world map or globe or an image of the earth. From space the biggest expanse of colour you will see is ? Yes, blue. These are the mighty oceans and seas which can be as deep as 11,000 metres – the Pacific, the Atlantic, the Indian Ocean.

Q: Can you think of anymore?

So, you might be wondering why we are worried about water and water shortages at all, when so much of our planet is covered in the stuff! Well, here is the reason why.

The problem that we have is that only 2% of the world's water is fresh water, and therefore suitable for us to drink. Most of that fresh water is locked up as ice at the North and South Pole and in glaciers. Some more is stored underground in aquifers and wells. That leaves only a very small amount in lakes and rivers that can be easily accessed for our everyday needs.

Added to that is the problem of water pollution and the fact that we have to treat and clean our water before it can be re-used. We also have to consider how many people live on our planet and that's a lot of people who need water every day just to stay alive. Hopefully, we are now beginning to understand why water, in some places, is a very scarce resource and why conservation is just so vitally important. Remember, there are also places on earth where it does not rain for months at a time.

We have already thought about how we use water each day and how lucky we are to have clean water piped to our homes. We could be forgiven for taking it for granted that we can just turn on the tap and clean, fresh water will come out which is also safe to drink. But people living in other parts of the world are not so lucky.

I would like you to imagine for a minute that you do not live in this country, and reflect on what it would be like if you had to survive on only a few litres of water a day. This is the reality for many people living in less economically developed countries. Millions of people lack sufficient access to water and sanitation to meet their basic needs. Many people die each year because they have contracted a disease from dirty water. Women and children walk miles every day to collect heavy loads of water that they carry home only to find that the water really isn't that safe to drink. Even still, some mothers have no choice but to give their children dirty water to drink. There are many organisations that are working to find ways to provide safe and clean drinking water for these people but it is not always an easy task. Some less economically developed countries lack the money to invest in digging wells and laying water pipes and have to rely on charities and foreign aid to help.

MAIN POINT 3 – CONSERVING WATER

In the UK, we do have times when water supplies are shorter than usual. I am sure you can all remember hearing news of a hosepipe ban or stories of dry reservoirs in the summer because our rainfall has been less that it should be. Fortunately we do not suffer water shortages like some of the people in other parts of the word.

So, do we really need to worry about water conservation at all? The answer is 'yes' - of course we do. Because not only are some parts of our country drier than others, but we also have times of water shortage here in the UK. But most important of all is understanding that water is a *global* resource. It doesn't just belong to us because we have better access to it. It is there for *all* humanity. Therefore it is very important that we do our best to conserve it and use it in a sustainable way so it is available for *all* people to use – not just now but in the future too.

I have used two important words there that you may have already heard in your Geography lessons but let's think for a minute about what they mean. 'Global conservation' means *not* to waste valuable resources, but to preserve them. When you use something in a sustainable way it means that you use it to meet the needs of the present but without damaging the resource for future generations.

So how can we conserve water and use it in a sustainable way? There are many examples of ways in which we can reduce the water that we use – here are just a few simple ideas for you to share with your parents and try at home:

- Turn off taps – for example, whilst you are brushing your teeth
- Have fewer baths, or take showers instead of baths
- Reuse your bath water for washing the car
- Only use your washing machine when you have a full load
- Collect rainwater outside for watering the garden

Q: Can we think of any more? *(Optional open question)*

SUMMARY

We have seen today that although there may be lots of water in the world, the truth is that we have limited supplies for us to use. The reality for some people in other countries is very hard indeed – they live in a world where their very existence is threatened due to lack of clean water. However, we have also learned that things *can* be done about this situation, and there are lots of things that *we* can do in our own homes, to make a difference.

CLOSING

On the 22nd March it is the UN World Water Day. This day helps us to think about the importance of working together to conserve our precious water supplies.

Every one of us can make a difference to help with the problems of world water supply. Think about how much water we would save if we all had a shower instead of a bath every day or turned the taps off whilst we were brushing our teeth or collected water to use on the garden. It would soon add up to a huge amount. If we all take simple steps such as these and make small changes to the way we use water, together it will make a big difference. So come on, you can do it! Let's get water wise!

Thank you for listening.

14

GOING FOR GOLD

AIM: Using the recent example of the London 2012 Olympics, this assembly considers success and the qualities that are needed to achieve it. It asks students to consider how this applies to their own lives and future ambitions.

Time required 5 to 10 minutes

HOOK

The stage is set. The venue is ready. We wait patiently for the main event. The place is London. The time is July 2012. All eyes are on our country. What is about to unfold is a rare event which, even in years to come, many people will remember. July 2012 was a time of celebration, a time of real achievement, a time when records were broken and medals were won. It was a special time, a great time – a time of true success.

OVERVIEW

Today we are going to think a little more about the success of Team GB in the London 2012 Olympics. I want you to consider the qualities that those athletes had that allowed them to succeed. We are also going to take some time to reflect on your success – what you hope for and what you need to do to ensure that you achieve it.

MAIN POINT 1: THE SECRET OF OUR OLYMPIC SUCCESS

The TV and newspapers tell the story. The medals come rolling in. People become hooked on the headlines. Team GB move up the medal table. Less well known athletes take centre stage. And then it is over, and Team GB have come an unbelievable third in the medal tables. They have been beaten only by the USA and China. 65 medals, 29 of which are gold and the nation is celebrating. The feel good factor is tangible. Our country is truly enjoying the 'sweet smell of success'.

However, we all know that this success did not just happen by luck or chance. It was not just a good day for the athletes or the result of 'a following wind'. Let's take some time now to think about what was it that made these athletes so successful?

Their success was the result of vision; ambition; years of hard work; training; perseverance; dedication and a determination to win medals, sometimes even against the odds.

I want to share with you some quotes from our medal winners that give us some clues to understanding the secret of their success.

> *"I can't believe I've done it. I had to stay focused and do what I had to do."*
> Jessica Ennis - gold medal winner, heptathlon

> *"We are such an amazing team. We could not have done it without each other."*
> Dani King, Laura Trott and Joanna Roswell - team pursuit

> *"I think I was the only one who believed I could win."*
> Greg Rutherford - long jump

> *"This is the best moment of my life."*
> Mo Farah - long distance runner

So, what do we learn from these quotes? What do they tell us about achieving success? Well, they tell us about the importance of staying focused on the task in hand. They show us that success is not just a solo effort but that, at times, we need to have support from others and work as a team. They teach us the importance of self belief even when others may doubt you. They also show us the great satisfaction that success brings – as Mo Farah says – it was the best moment of his life

Lots of other famous people through history have also commented on success. Here are a couple more ideas for you to think about. Theodore Roosevelt, the 26th president of the USA between 1901 and 1909, told us; *"Believe you can, and you are half way there."*

Vince Lombardi, an American football coach in the 1960s, famously said; *"The only place 'success' comes before 'work' is in the dictionary."* These comments reinforce what we have already learned about success from London 2012 – hard work and self belief are key.

Some people have succeed in sport as we have just seen, but others are successful in other fields. Some may be inspirational in business, others in politics, some as world leaders; others as inventors; some as humanitarians – whatever their field, they all share similar qualities.

Let's think about what these qualities are and how they relate to you in your approach to daily life. Do you have vision and ambition? Are you determined? Are you willing to commit to hard work? Will you persevere even against the odds?

Q: Can anyone name some other people in history who exemplified these character traits?

(Ask the assembly for examples of people here or great achievements – for example, the first four minute mile, the first person to reach the South Pole; to climb Everest etc.)

MAIN POINT 2: VISION AND PLANNING

The first thing is to have a clear vision of what it is you want to achieve. The athletes we talked about at the beginning of the assembly will have set their sights on a gold medal many years before they actually achieved it. They would also have a clear plan about what they needed to do. How can that apply to you? Take your GCSE or A level examinations as an example of a goal that you are working towards. You can set your sights not on a gold medal but on the best possible grades you can achieve. You can work out a clear plan too – it will not be a strict diet and training regime, but it will still be vitally important for you to know what you need to do and by when. For example, make sure all your work is up to date. Know when your examinations are taking place. Plan a revision timetable. Look up useful internet sites to help you. Work out the best way to revise and make sure you have completed all your revision in time. These are just a few ways in which you can ensure that you plan properly for your future success.

You also need to realise that success is not always a solo effort but rather the result of teamwork. Think of the Olympic gold medals that were won as a team? Consider all the people that will have supported the athletes to realise their dream – parents; friends; their coaches; sponsors; special trainers; dieticians......the list could go on. The same is true of striving for *your* success – it won't be something that you do alone. So ask for help and use the support that is there for you. You might ask a teacher for help with a difficult topic, go to a parent for support with a tricky situation or ask a friend to become a revision buddy. Whoever you choose, remember that support will help you to achieve success.

There are also times when achieving success is not easy and there are times that you may have setbacks along the way. At these times, you need to pick yourself, perhaps back up a step or two, and make an even more determined effort. Failure is an important part of success too - as long as we can learn from our mistakes and use them to move forward in a positive way, staying focused on the goals that you want to achieve. A popular popular quote that is often seen on posters and pictures in shops today says;

> *" when the storm comes, don't wait for it to pass - but learn how to dance in the rain."*

Q: What key message can we take from that quote that might change how we approach our lives today? *(Optional open question)*

Finally, one of the most important things in achieving success is always believing in yourself even when others around you doubt you. Tell yourself you *can*, and you are far more likely to succeed.

SUMMARY:

We have seen today that success is not just a matter of chance or luck. We have learned that you need to set your sights high and have a clear goal to aim for. Couple this with hard work, determination and self belief and you will greatly improve your chances of success in realising your hopes and ambitions.

CLOSING:

Wouldn't life be easy if you could just go into a supermarket and pick up a can of 'success' off the shelf? Imagine if all you had to do was open it up and there were all of your hopes and dreams achieved. You might wonder what I am talking about? How ridiculous – buying a can of success!! Right?

Well remember; success does come in 'cans' and not in 'can'ts'!!

Not only do you need this self-belief but you also need hard work to ensure that you make the most of what life offers you. I shall leave you with a quote from Thomas Edison as our final thought for today:

> *"Opportunity is missed because it is dressed in overalls and looks like work!"*

So, make sure you don't miss the opportunities to realise your dreams.

Rise to the challenge, work hard and believe in yourself!

Good luck, and thank you for listening.

15

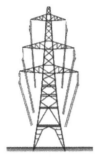 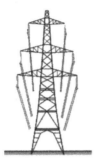 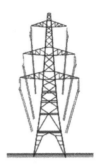

THE NATIONAL GRID

AIM: To inform students about the construction, purpose and management of the UK electrical grid and its impact on daily life.
Time: 8 to 10 minutes

HOOK

What would you do if you went home and had no electricity?

OVERVIEW

We rely on electricity everyday to conduct our lives, but do you know how the power generated by the power stations, wind turbines, solar panels, hydroelectric stations etcetera, is sent around the UK and into your home?

Electricity is something we cannot live without in our modern world. We have become so dependent on it in our daily lives that the thought of coming home and not being able to switch the lights on, browse the internet or turn on the television for a long period of time is almost unthinkable.

However, what many of you might not realise is that electricity in individual households is a relatively recent development. In fact, if some of you were to ask your grandparents or great grandparents a few of them might remember not having it at all.

To put into perspective how recently people in the UK lived without electricity; in 1920 only 6% of British households had electricity and those who did were some of the wealthiest in society at the time.

To fix this problem, it was decided by the government in the 1920's that an electrical grid should be built, which would distribute and provide power to the population of the United Kingdom. It was not until September 1933 however, that the original electrical grid was finished, and it would not be until 1968 that 90% of British homes had power flowing into them from the grid.

Many of you will have seen part of the electrical grid around your local area, normally in the form of huge electrical pylons supporting massive cables going across the countryside for miles into the distance. These cables are part of over 16,497 miles worth of power lines used to send electricity from power stations, wind turbines and solar panels to the majority of households in the United Kingdom.

The electrical grid, now known as the National Grid, is a monumental engineering achievement, which spans for miles across the entire length of the country, but most of us never even consider the work it took to build it, and the work it now takes to maintain, run, manage and repair it in order to keep the lights on in our homes.

MAIN POINT TWO – HOW IS THE GRID MANAGED AND MAINTAINED?

Maintaining and running the National Grid is no easy task, being comprised of such a massive infrastructure of major power stations, sub stations, electrical pylons and cables. Workers who maintain the pylons for example, have to climb the huge electrical towers all over the UK to make repairs. The regular size of the electrical towers is over 45 meters (150ft) high, so the job is not for someone who is afraid of heights. When workers are working on the towers one side is normally deactivated while the other side can still have up to 275,000 volts of electricity running through it. That vast amount of electricity would not just give you a shock, it is enough to hurl a person through the air and kill them instantly or to blow their arms and legs off. So a silly mistake could easily cost a worker's life. For these workers maintaining the grid is not just a difficult job, it is quite literally a life-threatening job. Nevertheless, it's a job that has to be done in order to ensure that the electricity we need to run our society is available to everyone who needs it.

It is not just the electrical tower workers however, who have a role to play in ensuring the grid is maintained. In a secret location in the UK, at a control centre, the National Grid is managed. Everything—from the distribution of electricity to ensuring that enough power is being generated to meet the demand of the UK at any given time—is controlled from that secret location.

Imagine for example, how the amount of power needed at different times of the day, varies. Think about the level of management it takes to ensure there is enough power available to meet the demands of the population.

Consider that everyone normally starts their morning with a cup of tea or coffee; this means around 50 million kettles are potentially boiling and all within a two-hour period. Boiling all those kettles in the morning takes a lot of electricity; a lot more than is needed by the population at four in the morning, when the majority of UK residents are asleep.

Some of you might be thinking at this point, *"But how do they reduce or increase the amount of power available, as you can't simply build a new power station or switch a power station off-or-on whenever you need more or less electricity?"*

Well, the answer is quite ingenious. During the early hours of the morning when most people in the country are in bed and much of the available power being generated is not being used, the excess electricity is sent to power huge pumps in Wales that send massive amounts of water from the bottom of a reservoir to the top of a reservoir. When extra power is required later in the day, the water is released from the top reservoir and then—thanks to gravity—used to drive huge turbines that produce electricity. The electricity produced from these Pump Storage Hydroelectric Power Stations can deliver a huge boost of power to the National Grid, when extra capacity (electricity) is needed. These turbine-powered stations can be switched on-and-off very easily and can deliver the extra power required almost instantly to the grid.

As an example, think about the power requirements after a popular television program or sporting event has finished being broadcast, when millions of people in the UK normally switch on their kettles to make a hot drink. If you asked a coal power station to deliver the extra energy needed to power all those kettles, it could take them around 30 minutes to increase the station's output to provide the extra energy needed. You might laugh at this point, thinking that all this sounds a bit ridiculous, but the extra power gained from sending huge amounts of water down a reservoir to turn massive generators often helps provide the extra electrical energy the nation needs when the demand for electricity is at its highest.

CLOSING

So remember, the next time you switch on a light, boil a kettle or see those huge pylons across the countryside, think about the time, effort and energy it takes to build, maintain and manage the National Grid and what life might be like without it.

Thank you for listening.

16

TRANSATLANTIC CABLE

AIM: To inform students about the construction and use of early transatlantic communication cables.

Time: approximately 10 minutes

HOOK

How long do you think it took to send a message from England to America by boat in the early 1800s?

OVERVIEW

We all use technologies such as mobile phones and the internet to communicate with people all over the world and never give it a second thought. Many of us who use technologies such as Skype or FaceTime for low cost, often communicating for free with people thousands of miles away and think nothing of it. It was not that long ago, however, when two-way communication with other countries separated by vast oceans was not possible, and people had to send handwritten letters instead, which often took weeks or months to arrive at their destination.

MAIN POINT ONE – THE FIRST TRANSATLANTIC CABLE

We often forget that 20 years ago, the thought of being able to see a family member on the other side of the world, through their own personal camera and to be able talk to them for free, in real time, was unthinkable. If you ask your parents or grandparents if they ever imagined something like Skype or the Internet would exist when they were your age, they will probably say no.

Some of the earliest forms of electronic communication, such as the Telegraph, were very primitive compared with today's technologies and could only send messages in Morse Code (a series of 'on' and 'off' tones) requiring skilled operators to encode the message in one location and then decode it in another. The telegraph and Morse code was used between the 1800s and early 1900s to communicate with people over vast distances.

The invention of the Telegraph and 'Morse Code' revolutionised communication on land and over some small sea distances in the early 1800s, but electronic communication over large distances separated by oceans was still not possible. Getting a letter from London to New York, for example, took a minimum of ten days by sail-powered ship.

However, in the early 1850s a successful American business man named Cyrus Field decided that an undersea telegraph cable should be laid between America and Europe. The planned cable would start at Newfoundland in Canada and end in Ireland, stretching a distance of nearly 2,000 miles across the Atlantic. Although wealthy, Field knew that the cost of laying this cable would be too much for him to meet alone, so in March 1854 Cyrus Field persuaded a few other successful businessmen to help him pay for the transatlantic cable. Together they pledged $1,500,000.00 to the project (about $42,900,000.00 in today's money) and founded a small corporation called, not surprisingly, *The New York, Newfoundland and London Telegraph Company.*

The project did not start well. Much of the original money invested into the project by Field and his associates was used sending a cable over the 60-mile Cabot Strait, which separates Newfoundland from mainland Canada, and then laying the cable through vast forests and over mountains to reach the eastern shore of Newfoundland itself. In autumn of 1855 the cable had only reached Newfoundland's Eastern shore and still had nearly 2,000 miles of ocean to cross to reach Ireland.

With little money left for the project, Cyrus Field decided to travel to London in 1856 to persuade more investors to help fund the completion of the cable. He went to see some of the wealthiest business persons, Lords, landowners and others to sell shares and thereby raise the money he needed for the project.

Mr. Field hired an English doctor with the eminently memorable name of 'Edward Orange Wildman Whitehouse'. Whitehouse was not only a surgeon, but also a telegraph enthusiast. Hired as chief electrician for the project, Whitehouse assured Cyrus Field that the cable would work. However, many other engineers of the 1850s doubted that the electrical current could even travel the vast distances required across the Atlantic. Field however, was determined to get the cable laid and listened to Dr Whitehouse instead.

After securing investors in England, Cyrus Field went back to the United States and worked to persuade the United States Congress to help with the cable. Eventually they agreed and the United States Navy lent a ship—the *USS Niagara*—to help lay half of the cable. The British government also provided a ship—the *HMS Agamemnon*—to lay the other half of the cable, because no single ship existed in the 1850s that could carry a 2000 mile long cable. After multiple failed attempts in 1857 and 1858 to lay the cable, the ships met on July 29, 1858, in the middle of the Atlantic Ocean, and the cables on the USS Niagara and the HMS Agamemnon were joined together.

After the cables were joined both ships set off in opposite directions.

The HMS Agamemnon headed for Ireland and the USS Niagara headed for Canada. Test signals were sent from both ships every hour or so to ensure that communication over the cable was working. Eventually both ships arrived successfully at their respective destinations, bringing their ends of the transatlantic cable ashore. The first message was then sent over the cable, stating,

> *"Europe and America are united by telegraphy. Glory to God in the highest; on earth, peace and good will toward men."*

Once the first message had been sent Dr Whitehouse sent another message, written by Queen Victoria to the President of the United States James Buchanan, explaining her hope that the communications cable would allow for,

> *"..an additional link between the nations whose friendship is founded on their common interest and reciprocal esteem."*

MAIN POINT TWO – PROBLEMS WITH THE FIRST CABLE AND THE LAYING OF THE NEW

Once the cable had been laid and two-way communication had been established between England and America, people were rejoicing in the streets. Extra issues of newspapers were printed celebrating the achievement and street parties were held. Those working with the cable however, started to realise that something was wrong.

The queen's short message had taken around sixteen hours to successfully send - which was far longer than it should have taken - and as time passed messages were taking longer and longer to send and receive. Dr Whitehouse attempted to correct this problem by applying very high voltages to the cable at the Irish end, although by his actions the insulation of the cable was damaged somewhere along its 2000 mile length, as it was not designed to accommodate such high voltages and within three weeks the cable no longer worked.

Cyrus Field was called to an enquiry in London to determine what had gone wrong and Dr Whitehouse was officially blamed for the failure through his actions of sending too high voltages down the cable. Field's Company was criticised also for hiring an electrical engineer with no recognised qualifications in the field for the project.

Many of you at this point might expect that Cyrus Field gave up on the project, but you would be wrong. Instead he replaced Dr Whitehouse with Professor William Thomson, who later became known as Lord Kelvin. Thanks to one of his creations called the *Kelvin's Mirror Galvanometer* - and along with advances in submerged cable design - a more suitable cable was created and the transatlantic project was resumed in 1865.

The biggest ship in the world in the 1860s was a British ship called the *SS Great Eastern*. It was designed by the famous British engineer Isambard Kingdom Brunel and was purchased by Field and his company to lay the new cable.

Owing to the ship's vast size, it was able to hold the 2000 miles (7,000 tons) worth of cable for the journey between Ireland and Newfoundland. This meant that only one ship would be needed to lay the new cable - thus reducing much of the costs and the complications of the first attempt. On the 27 July 1866, the newly laid and fully operational transatlantic cable was pulled ashore in Newfoundland at the village of Heart's Content. Since that day direct communication between England and America has not been broken.

CLOSING

Many of us today take instantaneous worldwide communications for granted, and we often forget that not so long ago this communication was not possible. It is thanks to pioneers such as Cyrus Field and many others throughout history, that we live in a world today where we can communicate over vast distances - instantaneously.

17

"THE THREE RRRs!!"

AIM: To encourage an awareness of the importance of conserving resources, this assembly considers how we are a wasteful society and then goes on to consider how recycling, reducing and re-using can help us to use our resources in a more sustainable way.

Optional Props: Powerpoint slides, photographs or examples of waste.

Time: 10-15 minutes approximately

HOOK

If I spoke to your grandparents and perhaps even your parents and asked them what they thought about when I said 'the 3 Rs', they would probably come up with a very different answer to the one that you are likely to give me. To older people, 'the 3 Rs' are *'**R**eading, w**R**iting and a**R**ithmetic'* – the three key skills that were so important for them to learn at school. However, times have changed. Whilst it is still very important to learn these key skills, we have come to associate this 'three Rs' acronym with something quite different.

Who can tell me what else might come to mind when I talk about the 3 Rs today? (*ask for responses from the audience at this point*)

Yes, that's right. We all know that today we are talking about

Reducing, **R**eusing and **R**ecycling the things that we don't want or need anymore.

OVERVIEW

Today I am going to ask you to take a few moments to think about the things that we throw away every day. I am going to ask you to consider ways in which you and the communities in which you live can reduce the amount of waste we produce. We are going to think about some practical strategies to reduce our waste and how we can reduce, reuse and recycle for ourselves.

MAIN POINT 1 – OUR THROWAWAY SOCIETY

'Waste' means those things we throw away or cannot use any more. We dispose of a huge amount of waste every day. Just think about what we put in our bins at home, at school, or on the street every day and then multiply that for the 63 million people that live in the UK today – that is a lot of waste. But just how much waste is it?

Here are some fantastic facts for you. Estimates tell us that the UK produces more than 100 million tonnes of waste every year. What does that amount of waste look like? Hard to picture? Imagine an African elephant that you have seen on TV or in a zoo. An adult male weighs about 7 tonnes. This means that we throw away in weight each year the equivalent of over 14 million African elephants!

So where does all that waste go? Another fantastic fact for you is that the average person throws away their whole body weight in rubbish every seven weeks. Collectively, this is enough waste to fill the Royal Albert Hall with rubbish every two hours.

Let's have a look at some of the things that we might throw away. *(the presenter may display some examples of things that have been thrown away at school or from a home in a typical week - or, just field suggestions from the audience and list them for later reference)*

Have you ever wondered what happens to that waste once it has been thrown away? Where does it go after the black bins have been put in the refuse truck or the wheelie bins emptied? Where do you put the equivalent of 14 million African elephants? Let's have a think for a minute - about what happens to our waste, and why it is important that we start to reduce the amount that we all produce.

The first place that our waste might end up is in a landfill site. Not only does landfill look unsightly and takes up large amounts of room but there are concerns about pollution caused by landfill too. We still use landfill because we produce too much waste and landfill is a convenient solution. But did you know that landfill not only contaminates underground water supplies but also releases a gas called methane.

Methane is a very powerful gas formed from decaying rubbish which contributes to global warming. We all know that global warming will cause temperatures to rise and that the knock-on effect could be the melting of the global ice caps and sharply rising sea levels, along with all the unsettling consequences of those changes.

Not only does waste come with an environmental cost, but imagine how much money it costs to pay people to collect our waste, process it, bury it in landfill, incinerate it.. and the list goes on. So, we can see that managing our waste in a sustainable way is not only good for the planet but also good for our pockets too.

MAIN POINT 2 – RECYCLING, REDUCING AND REUSING

When we recycle we make something new from our waste products. This is good because it saves money, saves energy, creates jobs and saves resources that would otherwise have been used to make new products. It also reduces the amount of waste that we have to put in landfill or send for incineration. As a country we only recycle 17% of our waste – this is quite low when you compare the figure with some other countries in Europe that recycle over 50% of their waste.

Just under two thirds of the rubbish we put in our bins could be recycled. Just think about what we are wasting – for example one recycled tin could save enough energy to power a TV for 3 hours.

Count up the tins you use each week and then work out the energy you save by recycling them – it might add up to quite a few episodes of the X Factor!

So... what can be recycled? Let's go back to the items we discussed earlier that we had thrown away. Could we have recycled any of those? Can anyone give me examples of things that we *can* recycle? *(If the presenter chooses not to use props, then this section can be done verbally.)*

Let's have a think about some of the things that we can recycle everyday: glass bottles; plastic containers; cans; newspapers; paper; batteries; light bulbs; cardboard. And what about water – that *also* gets recycled every day. The chances are that the water you are drinking has passed through another person at some point. Some people say that the water in London can have passed through seven different people before you might drink it! Lovely eh!? Next time you visit the supermarket, have a look at the recycling bins in store and also in the car parks for more ideas – this is a great place to take unwanted items that you cannot recycle from home.

We can also *reduce* the amount of waste we produce. Many companies have already taken this on board and are reducing the packaging they use for their products. Have you noticed how the packaging for your Easter Eggs is getting smaller each year or the increasing products that you can buy refills for? Instead of buying a new jar of coffee for example, you can now just buy a refill and save the need for another glass. There are many other ways to reduce the resources that you are using – and we will come back to this a little later in the assembly, but think for example, about the water that you would save each day if you had a shower instead of a bath or the petrol saved by walking to school rather than taking the car.

Another sustainable strategy is to *reuse* resources wherever possible. For example, you could take your unwanted toys and clothes to a charity shop so that another child can benefit from them again, or reuse envelopes or the back of old stationery as scrap pieces of paper - for those all-important homework projects. If something is broken, you could try to repair it rather than replace it. All of these will help to conserve resources by *reusing* rather than replacing.

MAIN POINT 3 – WHAT CAN *WE* DO?

Hopefully, you will now understand what it means to recycle, reduce and reuse when it comes to conserving our resources and reducing the amount of waste that we produce. Let's have a think now about some of the things that you can do at home and at school to make a difference. Perhaps you already have an Eco Council in school that can discuss this issue further and if not, maybe you would like to set one up. Here are some simple ideas for home and school:

- Have a paper recycling bin in each classroom
- Have recycling bins to sort break and lunch time waste in the canteen
- Sort your rubbish at home and either put out for recycling or take to the recycling bins at your local supermarket
- Design a poster to inform and encourage others to recycle more
- Donate your old clothes or shoes to charity
- Have a sale to pass on your old toys
- Try to reduce the amount of energy you use by not leaving appliances on standby or using energy efficient light bulbs
- Use email rather than printing if you can
- Start a compost heap in your garden
- Borrow things from friends rather than buying new if you can
- Buy products that don't have a lot of packaging.
- Refill printer cartridges
- Recycle your old mobile phones
- Reduce your water use by turning off taps, having showers, keeping a water butt to collect water for washing the car and watering the garden

There are many more things that you and your family and friends could do to help conserve our precious resources and reduce the problem of waste. Perhaps you could do a little research of your own and come up with some ideas that you can share in school. You only need to take a few small steps to start making a difference.

SUMMARY

We have seen today that we are still a wasteful and throwaway society. We have learned that this not only costs money to manage but can harm the environment too. We have seen that we can reduce the problem of waste by recycling, reusing and reducing our resources and that there are simple steps that we can all take to help.

CLOSING

Perhaps this week you can reflect on the things that you have thrown away. Have a think about the resources that you use every day. Ask yourself whether any of the things that you throw away can be recycled or reused. Can you reduce the amount of any of the resources that you are using? Perhaps you will decide to start by doing some simple recycling at home or by making sure that you always use the recycling bins in school.

Maybe you will reduce the resources you use by remembering to turn off the lights when you leave a room, or by only using the amount that you need. If you are already taking steps, can you share these with friends and encourage them to do the same?

Individually we can all start to make a difference – collectively, we can have a huge impact. Taking action now will benefit the environment, not only now, but also in the future. Remember, recycle today and we will still have resources tomorrow.

Let's protect those precious resources while we still can.

Thank you for listening.

18

HUMAN RIGHTS

AIM: To explain what human rights are.
Time required 5 to 10 minutes

<u>HOOK</u>

During the Second World War, about six million Jews were killed; that was about two thirds of the population of all European Jews.

Millions of other people were murdered including Romanian gypsies, homosexuals, people with disabilities and other political and religious opponents.

The total number of holocaust victims was between 11 million and 17 million people.

After the war a charter of human rights was set up by the world leaders of the time including Winston Churchill. This charter aimed to make sure that governments could not misuse their power and harm their citizens.

OVERVIEW

Human rights are all about freedom to choose, but what do we mean by freedom?

Well, ask yourself what is your favourite music or pop group. Do your friends agree with you that this is the best music? Some probably do but others may like different musicians. Let us imagine that the government decided you were no longer allowed to listen to your favourite music. How would you feel if they banned it?

Although it's only a very small part, freedom to have your own taste in music is part of your human rights and you would probably be very upset if these rights were infringed.

MAIN POINT 1: FREEDOM OF RELIGIOUS BELIEF

Do you believe in God? Should everyone believe in the same God or should people be able to choose their religion? Should people be allowed to not believe in God? In this country, we are free to choose our religion or to choose to be an atheist, an agnostic or any other type of non believer.

There are a number of countries in the world where this choice does not exist and where you can be executed for not believing in God. In 2014, these included Afghanistan, Iran, the Maldives, Mauritania, Pakistan, Saudi Arabia, and Sudan.

The reverse is true in North Korea where you are not allowed to believe in God. People who practise their religions can face imprisonment, torture and even execution.

In Quebec, Canada, legislation is being put through government to ban public employees from wearing any overt religious symbols including face veils, turbans and Jewish yarmulkes.

Human rights give you the choice of what to believe.

MAIN POINT 2

Should you be entitled to say what you think?

Yes you should, but there are limits. You cannot be, for example, racist, incite religious hatred or be Homophobic. However, within these limits you are allowed to express your opinion on anything, even if others disagree with you.

As the Italian Scientist Voltaire once said

'I may not agree with what you say, but I will defend to the death your right to say it.'

You also have a right to a private life.

This includes the right to choose your sexuality and not to be persecuted for being, for example, homosexual.

Discrimination on the grounds of sex, whether you are male or female you should be treated equally.

Respect for private and confidential information, particularly the storing and sharing of such information. How do you think you would feel if you had got the lowest mark in all your exams at school and someone published this information on the school notice board?

The right to control the dissemination of information about your private life including photographs taken covertly, for example, would it be right for someone to secretly take a photograph of you on the beach on holiday then publish it on a social media site.

Family life is also included the state cannot take you away from your parents without very good reasons.

You also have the right to meet people and gather together in a public place. Can you imagine what it would be like if you had arranged to meet your friends in the park and the police came along and told you that meeting together was not allowed?

The right to an education is also included, although some of you would perhaps prefer not to be in school, this is considered so important that education is part of your human rights.

When you reach the age of 18 you will be allowed to vote which gives you a say in how the country is run and is possibly one of the most important human rights of all.

MAIN POINT 3

These rights all seem to be common sense but the Human Rights Act can be controversial.

In July 2013 the terrorist suspect Abu Qatada went to Jordan to face trial. He was not charged with any offences in Britain and the evidence for his trial in Jordan may have been obtained by torture.

There was a huge media outcry that he was a dangerous terrorist and should be deported even if it was illegal to do so. Evidence obtained by torture is nearly always unreliable as people will say anything to stop them being tortured. Imagine you had been accused of something by evidence obtained by someone being tortured - you were innocent but were convicted to a long prison sentence on evidence that was not true.

The Human Rights Act prevented his deportation; he was not allowed to leave the country and stand trial in Jordan until judges were convinced that his trial would be fair and the torture evidence would not be used against him.

SUMMARY

To summarise, some of the rights provided by the convention are listed below.

- Freedom from torture and very cruel treatment (Article 3)
- The right to liberty and security (Article 5)
- The right to have your own thoughts, religion and beliefs (Article 9)
- The right to privacy and family life (Article 8)
- Freedom of expression and opinions (Article 10)
- The right to meet people and gather in public places (Article 11)
- The right to an education (Article 2 of Protocol 1)
- The right to vote in elections once you reach the voting age (Article 3 of Protocol 1)

These rights have been fought for over many years and they will only be there for as long as people defend them.

Let us think about some situations where human rights could be important.

Imagine there is a football competition and there are girls' teams and boys' teams playing. The prize is several thousand pounds. The grounds man said that the girls' teams were not allowed as much time to practise as the boys as their matches were not really important. He also asked them to use a changing room which had pictures of topless girls all over the walls.

I would like you to consider if their human rights have been abused - if so which ones?

At a council meeting all the councillors were asked to turn up at 8 pm so they could say prayers before the meeting. Councillors who did not want to say prayers and turned up after they had been said were put down as late for the meeting. This went on records which were available to their electorate.

Again do you think their human rights have been infringed if so which ones?

The answer to the first situation with the football team is that there was discrimination on grounds of sex.

The second situation about the council meeting was the right to choose your religious belief.

CLOSING

Human rights can often seem irrelevant or ridiculous and not all people agree that the Human Rights Act is a good idea. However they become vitally important to you if your human rights are infringed. We often do not realise how important something is until it has been taken away.

19

LEFT WING, RIGHT WING

AIM: To explain the difference between right wing and left wing views. (Suitable for times when elections are taking place)
Time required 5 to 8 minutes

HOOK

I am sure you have all heard the terms left wing and right wing applied to political parties. Generally the Conservative party is considered 'right wing' and the Labour party 'left wing' but what is meant by 'left' and 'right' wing? Is there anyone here who can tell me what these terms mean?

OVERVIEW

The following letters may give you an idea about what the views are of right wing and left wing people.

Right wing letter

"There are far too many people scrounging on the dole the welfare budget is out of control. Why should I pay my taxes for people to stay in bed all day? I mean there are some people who live opposite us and the curtains don't twitch in their house until 10 am when we have been at work for two hours.

They should be made to work to get their benefits . At least they would have to get out of bed and get some idea of the work ethic. Too many people think the country owes them a living. Youth unemployment is terrible; there are always loads of young people unemployed - they should have their benefits taken away, they would soon find work then."

Mr D Mail

Left Wing letter

"Most of the people on the dole have paid in money as part of their national insurance contributions. They have worked all their lives and are just claiming what is owed to them while they are temporarily unemployed. In fact, only 2% of people have been unemployed for more than 2 years, and even of that 2%, most have paid in more to the system than they will ever take out. It is not their fault that there is a recession and the companies they work for have gone bankrupt. In fact I blame the greedy bankers for causing the world recession.

If people are made to work to get their dole money then companies will be happy to employ them at no cost to themselves. The problem is that if a company is employing someone on the dole and not paying them, they will not employ anyone in a proper job where they pay them a wage. I mean they are thinking of making the road sweepers redundant where I live. It won't be a problem as they can just tell them they have to do the job for free to get their dole money. If people are made to earn their dole payments by working for private firms the taxpayer subsidises the profits of large companies and unemployment soars.

Yes there are always young people unemployed but it is not the same young people all the time. They leave school or college and have to find jobs. They then get older and are no longer part of the youth unemployment figures. Would you really like to see people with no benefits literally starving on the streets? This happens in other countries and is not something I would like to see in Britain.

Mr G Ardian

On the subject of the National Health Service

Right Wing views

"The health service is far too expensive, we just cannot afford it. Let us look at America where they have some of the best health care in the world. People take out private insurance and get brilliant treatment, they have rooms to themselves in hospital and Americans can treat many conditions that are not available on the health service.

If we got rid of the health service we could give everyone the money back in tax cuts and they would be much better off. They could use some of the money to buy health insurance. Competition in the market would bring prices down, making health care cheaper for everyone."

Mrs D Express

Left wing views

"Yes the health service in America is very good for those who have money. If you can afford to pay for health insurance then you get fabulous treatment. Unfortunately most of the American population cannot afford health insurance.

The privatised US system has the highest healthcare spending in the world. Privatised industries must make a profit for shareholders, that is why they are expensive.

Of all the money the US government takes in as revenue from taxes it spends 15% on health care and treats about 28% of the population.

Of all the money the UK takes in as taxes they spend 8% on health care and treat 100% of the population.

In America children have been sent home to die because their parents cannot afford health care - would we really want this situation to happen in Britain?

There must be competition in the system but this should be used to keep costs down - not to make huge profits. Some services should *not* be privatised; they should be there for the public good."

Miss D Mirror

MAIN POINT 1

These letters were made up and perhaps a little extreme in their ideas.

However, when you are 18 you will have the right to vote and it is important that you understand what you are voting for. The country always has limited resources and governments can never please everyone. Nevertheless, politics affects everything you do in life. If we don't vote we have no say in education, health, welfare, housing, the tax we pay, the money we earn etc, etc.

Extreme views in either direction can cause massive problems. In North Korea today we have a left wing extreme dictatorship - you can be executed for owning a copy of the Bible or watching South Korean entertainment.

The Nazis were a very right wing group in Germany; they murdered millions of people for their religious beliefs. Gypsies, the disabled and homosexuals were all killed. Anyone who was not blue eyed and part of the Aryan race was discriminated against.

It is a bit of a twee statement, but it is said that, *'for evil to triumph good people merely need to do nothing.'* Not voting in an election is doing nothing. In some countries such as Australia it is illegal not to vote. Do you think this is a good idea?

If a political party said they would make it illegal for men to vote as they tend to be violent and extreme would you vote for them?

It sounds silly, but women were not allowed to vote until 1928 in Britain. In France it was 1944 and in Lichtenstein, close to Switzerland, it was 1984. Lichtenstein has the largest GDP per person in the world and the lowest external debt and unemployment. It is a constitutional monarchy headed by a prince (this is not because they can't afford a king) do you think it was right wing views or left wing views that prevented women from voting in Lichtenstein?

SUMMARY

We have heard the different views of people from the right and the left - also that extreme right or left wing views usually lead to dictatorships and a total removal of human rights.

The importance of voting has been stated and the idea that if you don't vote you have no say in anything. How—by doing nothing—'evil' will triumph.

CLOSING

Information is really important when considering who to vote for. You need to understand what the parties stand for. Read about some of the policies of the current political parties and decide what you think is right and what you think is wrong. No one can tell you how to vote and many people have died in the past getting you the right to vote - so please use it.

Thank you for listening.

20

ROOM 101

AIM: To think about what students would really like to get rid of 'forever'.

Time 15 to 20 minutes

Note to presenter: *this assembly is <u>not</u> 'ready to go'. It does not follow the usual pattern of assemblies in this book. It is designed as a whole-group activity with pupils' voices as the central theme. This assembly provides an opportunity for students to discuss issues about their school that make them feel unhappy, and provides an opportunity for staff and students to consider ways in which these difficulties could be improved or resolved.*

START OF ASSEMBLY

Explain to the students that the idea of 'room 101' came from George Orwell's novel '1984'. 'Room 101' is a place where you would find the things you most fear or dislike. Winston Smith was a character in Orwell's book. He was tortured with rats in Room 101; his very worst fear.

The basic idea is to play a game that recreates the BBC's *Room 101* television series where contestants nominate pet hates they want eliminated from society. Each contestant puts forward the case for getting rid of his or her pet hate and, depending on the persuasiveness of the arguments, the presenter chooses whether or not it should be consigned to Room 101.

To play the game, choose one student to be the presenter and two students and a member of staff to be the contestants (the member of staff could be a plant who will put forward a common staff complaint; this can be particularly effective if the pet hate is something that both staff and students dislike, e.g. homework, lunch queues, outdated/smelly toilets). Ideally, have some objects to hand that can be thrown into Room 101.

Some examples of student pet hates are; school rules; teachers (although this one needs careful handling); uniform; school canteen (food, rules, queues); PE; assembly; tutor/form time; detention; tests; exams and targets; toilets; planners; reports; before/after school revision sessions; and parents' evenings.

This assembly can be even more powerful if the students themselves can organise and run it. Student leaders such as sixth formers, prefects and house captains are potential choices. However selecting students who are clearly not speaking on behalf of the school can result in an even greater impact (e.g. students who are not in leadership positions but who are often very vocal about their grievances). It may be useful to consult tutor groups and/or houses, as well as members of staff, about the best way to organise your assembly.

Some examples of student pet hates and the justification for sending them to Room 101 (props can be used as appropriate):

SCHOOL UNIFORM

Ties
Ties are hated by most students. Why do we have to wear objects that were invented for people on horseback to wear during cavalry charges? What relevance do they have today? People feel they make you look smart but the way most students wear them, they make us look more of a mess. Even most businessmen and world leaders have stopped wearing them. Ties are about as fashionable as a chemistry teacher's cardigan.

Ties are a serious health and safety hazard. They are often used by students to throttle each other. In practical lessons, ties are positively life-threatening. We're always told to tie our hair back or wear hair nets yet we still have ties dangling dangerously down in front of us. At any moment, we could be turned into a raging inferno as our swinging ties stray into the flame of a Bunsen burner or decapitated as our ties get caught up in a piece of DT machinery.

The one good thing about a tie is that you never have to bring a snack with you. The leftovers that have been deposited on your tie after a term's worth of school dinners mean there's always something to nibble on. Ties are rarely washed more than once in a millennium. They are a hygiene hazard worse than a salmonella sandwich.

SCHOOL SHOES

The rules on school shoes are designed to confuse us. Who decided we should still be wearing plain black shoes instead of, for example, the more modern and practical alternative of trainers? You tell us to get out into the fresh air and get some exercise and then moan at us when we wreck our school shoes playing football out on the field and get our skirts and trousers muddy because we've fallen over (which, incidentally, probably wouldn't have happened if we'd been wearing trainers in the first place). Trainers are comfortable, provide practical support for your feet and have cushioning that prevents knee and ankle injuries. Why are we encouraged to wear flimsy loafers and ballet pumps when there is a much more sensible alternative?

BLAZERS

Who in the name of all that's holy created blazers? They are expensive, uncomfortable and incredibly hot in the summer. Being British, you make us wear them even when it is 30 degrees in the shade - and then you wonder why we smell! Enough said on that one.

SCHOOL RULES

School rules are just discrimination on a massive scale. For example, the rules about jewellery and make-up are simply unfair. Why aren't we allowed to wear nail varnish or make up when they make us look really nice? Members of staff can wear make-up, nail varnish, high heels, jewellery and whatever else they want. Why can't we dress up too – especially when we get so many complaints about our shabby appearance?

STANDING

Standing behind your chair at the start of a lesson or standing up whenever an adult enters the room is a completely outdated waste of time. I have been in some lessons where we got no work done at all as there were so many visitors. We did more squats than in a PE lesson. How are we supposed to learn when we are doing an impression of a penguin on a trampoline?

HOMEWORK

Has no-one in education heard of the EU Working Time Directive? We work hard all day in lessons. Add the time for after-school sports and activities and bus journeys that can take up to two hours, and we are all putting in at least forty hours a week before we even start on homework. Those of us who do more activities or travel further to and from school are putting in up to fifty or sixty hours a week.

When you finally get home from school, you're so stressed with all the pressure that you would love to relax and phone your mates, play a game or even just go for a walk (we're always being told how important fresh air and exercise are). No chance - you have to be able to speak French like Napoleon by 7.15 and then derive Einstein's theory of relativity by half past eight. By ten o'clock, you've learnt the names of all the capitals in the world and by midnight, you have re-written and improved Romeo and Juliet. There's no time to eat, sleep or even go to the toilet - it's straight back to the grindstone.

The teachers hate homework just as much as the students. Have you ever actually seen a piece of marked homework? It's as rare as Unicorn do-do.

On a more serious note; according to the European Working Time Directive, we are not supposed to work more than 48 hours a week. But some of us do that much studying in just three or four days!

At weekends too, most of our 'free time' is spent doing homework - which easily puts us over the 48-hour limit. The situation is so serious that children as young as five are being diagnosed with stress. And if homework is so critical to success, then why do other European countries that set no homework at all have far more successful economies than us?

DURING THE ASSEMBLY

After each contestant has presented his or her pet hate, the students in the audience are asked to vote on whether or not it should go into Room 101. Before the voting, some audience members could also be given the opportunity to voice their opinions (students and members of staff).

All vote and the chosen object is either saved or sent to Room 101.

END OF ASSEMBLY

When the game is over, the presenter or a member of staff should elaborate, "This was a light-hearted exercise and really just for fun. However I would like you all to take a few minutes to think about things we really could put into Room 101 that would significantly improve your life at school. If you can identify changes that would be good for everyone, and explain how we could reasonably implement these changes, we would like to hear from you."

There should then be an opportunity either at the end of this assembly or soon afterwards during another assembly or form time, for some general discussion between students and teachers about how their school experience could be improved.

Notes:

21

ATHEISM

AIM: To explain what atheism is.

Time 5 to 10 minutes

HOOK

In the 2002 census, in response to the question, "What is your religion?" more people put down 'Jedi Knight' (the mythical religion from Star Wars) than Jewish. This wasn't because people actually believed they were Jedi Knights. It was a way of protesting because 'atheist' (someone who does not believe in God) was *not* an option on the census form. In other words, it was assumed that everyone must have a religion, when this is *not* actually the case.

OVERVIEW

Every day, without being aware of it, we are exposed to religious views. Every town and village has at least one church and often other places of worship as well. Prayers are regularly said on radio and television. Programmes such as *'Thought for the Day'* on Radio 4 and *'Songs of Praise'* on BBC1 all give air time to religions.

But surely, atheists have as much right to their beliefs as Christians, Jews, Hindus, Sikhs, Muslims and members of other religions - don't they? This assembly is about redressing the imbalance. It's a chance for us to find out what atheists believe.

We should bear in mind that everyone has the right to believe - or not to believe in God - as they see fit. This assembly will allow us to discover what atheists think about the world around us.

MAIN POINT 1: WHY DO PEOPLE BELIEVE IN GOD?

Before thinking about why people *don't* believe in God, it's worth considering why people *do* believe in God. 'Theism' means to believe in a god. Therefore, all religious people are 'theists'. 'Atheism' on the other hand is the non-belief in deities - or gods.

In the past, religion has often been used to explain the world around us. Thunder and lightning for example were the result of some god's anger. The crops were successful because God was pleased with us.

Religion also provides answers to deeper questions such as, "where did we come from?" (God created us). "Why are we here?" (to serve God) and, "what happens to us after we die?" (Heaven or Hell). The answers provided by religion make up the different belief systems.

But atheists point out that simply *believing* in something doesn't necessarily mean it is true, real, or a fact - otherwise, what happened to our childhood *beliefs* in Santa Claus and the tooth fairy? Atheists point out that there is no *scientific* evidence to support most of these religious explanations for the meaning of life, and in fact, a great deal of evidence to contradict them.

The answer to the question of why various people believe in a particular religion lies mainly in geography. Where we are born usually determines which God (or gods) we are going to believe in. People born into Christian communities tend to believe in Christianity. Those born into Jewish communities tend to believe in Judaism, and those born into Hindu or Muslim communities tend to believe in Hinduism or Islam respectively. The beliefs we are brought up with usually depend on our country of origin and the beliefs of the community around us.

MAIN POINT 2: RELIGIONS ARE TRANSITORY

A religion - any religion - has a beginning, middle and end. No religion is eternal. For example, the ancient Egyptians worshipped the God Amon Ra for 4,000 years - twice as long as Christianity has even been in existence. But today, Amon Ra does not have a single worshipper. That religion is at an end.

The Viking Thunder God Thor, is in a similar position. Even Zeus, the father of all Greek gods, no longer has any followers (apart from the ones on Twitter). These were extremely popular religions of their time, which have all now disappeared.

Richard Dawkins, a famous atheist who wrote a book called, *'The God Delusion'* looked at it another way. He said, "Most of us are atheists about most of the gods that have ever existed. I just go one God further."

Atheists argue that if God exists, surely religion should have existed from the time when humans first evolved? One interesting question they pose is why—when humans have existed as a species for at least 100,000 years—there is no evidence of religion and any belief in God until the last 6,000 years? If God really does exist, why did he or she wait so long to introduce religion to humans?

Most of the arguments about whether or not there is a God come down to the discussion of personal *beliefs* vs scientific *proofs,* and whether or not one supports the other. However, in order to be used as 'evidence' (in a scientific experiment for example) 'proof' has to be *measurable*, and the experiment or situation that provided that proof has to be *repeatable*. The big problem with the 'scientific proofs' used by religionists to support the existence of God, is that these particular proofs can't be measured or repeated.

Just declaring for example that the creation of the world *proves* God's existence doesn't actually provide scientific evidence that God is real. All that argument proves is that *the world* actually exists

- and not necessarily *how or why* it came to exist - right? On the other hand, there is a great deal of evidence to support *The Big Bang Theory* (that's the theory of the origin of the universe, not the television series).

A common argument used by religionists to explain the existence of God and to discount scientific approaches such as *The Big Bang Theory* is to point out that there are gaps in our scientific knowledge. This is true of course, and science researchers continue to work towards filling-in those gaps. This does *not* however, *prove* that God exists; only that we don't quite know everything yet.

Science does not know everything but we do know enough to put a man on the moon; to make drugs that actually cure diseases, and to make the millions of technological devices we use in everyday life; from cars to play stations.

To put this in context; would you use modern medicine to cure a relative of say tuberculosis or would you rather just pray that they get better? I know which option I would have more faith in. Tests have been carried out where people prayed for some of the patients in a hospital to get better, while the other half of the sample in the hospital was *not* prayed for. The results of that test showed that prayer made no difference whatsoever to recovery rates. However, when the patients were given the correct medicine they all recovered.

Evolution was often cited by religionists as a theory that cannot be proven, because there are missing links in the archaeological record. Those 'missing links' however, are no longer missing. Evolution made one phenomenal prediction suggesting that all life is related.

It is astonishing that mapping the genomes of living creatures has shown us that we are genetically 99.8% identical to chimpanzees. This proves that we evolved from creatures that were a common ancestor of humans and apes. In fact we are all 50% identical to a cabbage genetically! Many scientists would now like to see evolution moved from being 'a theory' to being an accepted scientific fact.

A question often asked of atheists is, "Do you have a moral code? Without religious guidance can you actually live a decent life?"

Well, it seems that humans do not conform to a moral code just because of religion. There are some very good Darwinian reasons why humans as a species are more likely to survive if they work with each other rather than against each other. Most people have perfectly good moral codes whether they are religious or not, and this has been the case throughout human history. Laws are obeyed because it makes a better society - not because people fear eternal damnation

Some interesting research has indicated that atheists are, on average, more law-abiding than those with religious beliefs. In fact, atheists supply less than 1% of the current prison population.

Statistics show that during 10 years in *Sing Sing* prison in America, those executed for murder were 65% Catholics, 26% Protestants 6% Hebrew, 2% pagan and less than one-third-of-one-percent, atheist.

SUMMARY

We have seen that religions in the past have had a beginning, a middle and an end, suggesting that today's religions too will be replaced in the same way in the future. We have also seen how religion is often based on culture or geography. Given that the genetic make up of chimpanzees is almost identical to that of humans, we have also raised the question of when and why religion came to exist? We have also established that even though atheists do not believe in God, that they still have a moral code and seem to commit less crime than those belonging to religious groups.

CLOSING

Remember that a personal *belief* is just that; it is *personal* and it is a *belief.* What you chose to *believe* is up to you of course. But it is always wise to look at both sides of the discussion; to evaluate the evidence scientifically; and not just believe things, unquestioned.

22

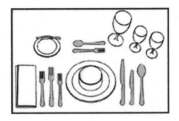

MANNERS

AIM: To understand why manners are important but also to understand they can be used as a way to divide people.
Time about 5 minutes

HOOK

What do we mean by someone who has 'good manners'? Why should we care? Does it really matter if you know your desert spoon from your soup spoon?

OVERVIEW

One day, sitting at the dinner table with what I hoped would be my future in-laws, I was asked, "Do you know the Bishop of Norwich?" I had no idea what they were talking about and so replied, "No, should I?"
"He's a very nice chap but he always forgets to pass the port!" was the answer. This was followed by some poorly-concealed snickering - apparently at my expense. I had no idea at the time that what had just happened would mean that I could never marry into this family.

Q: Does anyone know why?

MAIN POINT 1: WHAT ARE MANNERS?

A: According to the laws of etiquette, when drinking port at dinner, good manners require you to pass the port to the person on your left and to keep it moving. But I didn't know this. When the port arrived at me, everyone else already had some in their glass so I just put the bottle back down on the table. Why should it matter so much which way the port is passed or if it's put on the table where everyone can reach it? Is a lack of knowledge of this obscure tradition really a sign of bad manners and a good enough reason to ostracise someone?

Here is another story about manners. Whilst studying at the University of Cambridge, a young Tanzanian woman was invited to a charity dinner given by a local lord and lady. These dinners were renowned for the large quantities of lavish food served in many courses. The table was always set with hundreds of pieces of cutlery and many of the guests prided themselves on knowing exactly which implement to use and when to use it. As the young woman sat down, she picked up a finger bowl and drank from it. Finger bowls are used for cleaning your fingers between courses and are not meant to be used for drinking. Some guests began to snigger and others stared in horror. The lady of the house had seen the incident and, in a display of extraordinarily good manners, picked up her own finger bowl and drank from it as well. The other guests followed her example and all the sniggering and staring stopped.

MAIN POINT 2: WHAT'S THE POINT OF GOOD MANNERS?

Most people glaze over at the thought of a discussion or, even worse, an assembly, on manners. We tend to take manners for granted - until someone else's bad manners causes us a problem.

For example, you're on the bus on your way to school, chatting with your mates and minding your own business, when a business woman gets on the bus and pulls out her phone. "No, no! I told him that's not good enough. You get back onto him and tell him Wednesday's no good. I need it Monday. I mean it – don't take any other offers from him!" This wouldn't be so bad except that she's yelling down her phone so loudly that everyone else on the bus has stopped their own conversations (out of politeness, of course) to listen to her rant.

At last the phone call is over, everyone breathes a sigh of relief, you start chatting with your mates again and then... she makes another call! "I told you Monday – why did you tell him Wednesday?!!!" And on it goes. Sitting there, you think to yourself, "This woman is so thoughtless and annoying. What bad manners!"

Q: So what exactly *are* manners?

A: The dictionary says that manners are an agreed way of behaving or doing things, and we should all try to be aware of what is, or is not, socially acceptable. Things such as *not* shouting down the phone when you're on a bus; or *not* pushing people out of the way when we walk along the corridor and, most importantly, *not* being disgusting when we eat.

Manners are, first and foremost, social rules based on courtesy and consideration of others. They are supposed to make it easier for us to live with each other. Observing obscure dining traditions handed down from the aristocracy is not necessarily 'good manners' - not when it is done at someone else's expense. Good manners begin with a genuine consideration of others. They are *not* about making people feel uncomfortable when they break some obscure rule just because they didn't know about it - or couldn't know about it. In fact, it would be an extraordinary display of *bad* manners to publicly embarrass someone in situations like this. On the other hand, if you know you are going to be in unfamiliar surroundings it is probably best to find out in advance what is, or is not, considered 'acceptable'.

SUMMARY

So, manners are important, but not just because they're what someone else tells you to do. Manners are important so that everyone knows what to expect and how to behave.

CLOSING

I don't know about you, but I really don't care what the people sitting next to me on the bus are wearing (green hair, nose piercings and Doc Martins are fine by me) but I am always extremely grateful if they can sit there quietly without blasting their bad taste in music in my ear, shouting down their phones or blowing smoke in my face.

Manners are there to make it easier for us to live together and to get along better. Good manners are what stop you having to watch the person sitting next to you in the canteen picking their nose while you try to enjoy a slice of pizza... (it's just snot on).

I now thank you - very politely of course - for listening today, and ask you all to observe good manners as you exit the assembly.

23

A SHRINKING WORLD?

AIM: This assembly asks pupils to think about the map of the world and how it has grown in history. However, it also asks them to consider why people now say that the world is shrinking, linking this to the benefits brought to us all by globalisation.

Optional props: map of the world or globe.

Time: 10-15 minutes approximately

HOOK

Take a look at this map of the world (*show a map or globe to the assembly*). This is a well known image that we are all familiar with. However, has our world map always looked the same?

I wonder what this map might have looked if I was holding it up to a group of children 600 years ago? (*Take some suggestions at this point*)

Well, it would not include America – this was not discovered by Christopher Columbus until 1492 when he arrived in the Bahamas en route to Asia! It would also not include Australia which was not discovered until 1770 by Captain James Cook. At this time too, nobody knew of the mighty southern continent of Antarctica which was only spotted in 1772 and not explored until later. Therefore, our world has grown not literally but in terms of what we know of it throughout history as explorers have blazed trails and discovered new worlds.

OVERVIEW

Today we are going to consider how our world grew as explorers through history risked their lives to discover new worlds. We are also going to think about how our world, although physically not getting any smaller, is in fact 'shrinking' due to us being able to travel to faraway places and communicate worldwide more quickly than ever before. We will also reflect on how this helps us every day but also look at some of the problems this can bring.

MAIN POINT 1 - OUR CHANGING WORLD

We have already thought about how the world grew through history as explorers discovered new worlds and what the map would look like without the Americas, Australia and Antarctica. What if Sir Walter Raleigh had not set sail looking for gold and come back with potatoes! Imagine a world without crisps, fries, mash or chips. How strange that would be. What would we eat with our fish? What would go with our Sunday roast? How could we survive without salt and vinegar crisps?

However, Walter Raleigh also discovered tobacco. I think we would have been much better off if he had left that one in North America!!

So, our world got bigger as explorers added more places to the world map. I wonder if it would surprise you to know that the world today is still getting bigger in some places, but actually still shrinking in others? Let's take another look at our map to see if this is really true.

Q: Can you see two continents that look like pieces of a jigsaw that should fit together? (*Take some suggestions from the pupils at this point*)

A: The best example to look at is South America and Africa – see how their coastlines match? This suggests that South America and Africa were at one time joined together, but over time have been moving further away from each other thereby making the Atlantic Ocean bigger every year!

Does this mean then that our world is growing, even by a little bit each year? Well, the answer to that is no because some parts of our world are actually getting smaller. Take the Pacific Ocean – it is actually getting smaller each year as the Atlantic Ocean grows.

Q: So, what is this due to? (*Open up to suggestions from the pupils if you wish*)

A: This is all due to something called 'plate tectonics'. The crust of earth is like the thickness of the skin on an apple and is divided into huge sheets of solid rock called 'plates' (not the ones you eat your dinner off) that are floating on the molten rock below. These plates move a few centimetres each year either away from each other, towards each other, or side-to-side. The parts of these plates that are exposed above sea-level are what make up our continents.

This theory of plate-separation is the reason why we only find marsupials in Australia for example, because that continent separated from the main land mass and isolated those type of animals. This is also the reason why a new island, named Surtsey, suddenly appeared off the coast of Iceland a little over 50 years ago. Lava erupting from plates moving apart under the ocean had grown up as an underwater volcano and appeared, much to the amazement of local fishermen literally overnight, in 1963!

This amazing theory of 'plate tectonics' was first discovered by a German Scientist at the start of last century. Alfred Wegener proposed a theory of continental drift where he explained that the plates were being moved by huge convections. These currents in the mantle of rock and lava under the earth's surface were allowing continents to move thousands of miles apart over millions of years. This explained why the rocks on coastlines of different continents matched and why fossils of similar species were also found that could never have swam across the ocean to arrive on the other shore! However, this theory was not accepted for a further 50 years and it was only in the 1960's that Alfred Wegener's theory of continental drift was finally believed.

MAIN POINT 2 - GLOBALISATION

Q: Can anyone tell me how long it would take to travel from England to New York a hundred years ago by ship?

A: Well, it would have taken you about six days.

Now you can fly across the Atlantic in about six and a half hours! So, although London is only a few centimetres further away from New York than it was a hundred years ago, it is actually much closer in real travelling terms. This is because we have developed transport and communication systems that have made our world a smaller place. Man only took their first flight in 1903 but now there are over 43, 000 airports on earth. In 1969, we put our first man on the moon. How long will it be until we can take flights to the moon I wonder?

The solar system—and not just our planet—is shrinking too! Places that could only be reached by explorers are now accessible to all. For example, Scott lost his life trying to get to the South Pole in 1912 but now you can book a cruise to Antarctica. In the 1980s if you had a mobile phone it would be the size of a house brick and would rarely work. But soon we will be wearing our phones as a wristwatch! If you wanted to phone Australia a few years ago, you would have to book the call in advance and the connection would be terrible – today, you are connected via the phone, skype, internet or text within an instant!

These developments have brought us many benefits. Think about the products that you can buy in supermarkets all year round that are shipped to us from faraway places. Think about the products that we can buy made in other countries that take advantage of the latest technology and are now more affordable than ever before; cameras, phones, iPods, iPads, computers, cars etc ...the list goes on and on.

Dream of the destinations that you can now visit on your holidays, accessible in only a matter of hours. Globalisation has many good points and we are more inter-connected and inter-dependant as a world than ever before.

Like everything, though, we need to consider both sides. Here are some questions for you to consider.
- Does the environment suffer from the millions of miles clocked up by trucks, boats and planes each year, bringing this wide range of food to our shops?
- How is the atmosphere being polluted by the millions of aviation miles travelled each year on our holidays?
- How many people accept low wages and work in poor conditions to make our latest designer goods?

Globalisation – a good or a bad thing? Well, you could argue it either way. Perhaps, you might decide to do a little more research and become more informed – then you can make up your own mind.

MAIN POINT 3: OUR RESPONSIBILITY TO EACH OTHER

Well we said at the beginning of this assembly that the world had grown. Then we said that it was growing in some places and shrinking in others. Then, just to confuse you, we said that it was shrinking in real terms because of globalisation. So, now I want to really confuse you and tell you that our world is growing, growing fast, indeed growing too fast. Some would say at an alarming rate!

Here are some examples for you to think about and to reflect on what this might mean for our planet, our country and for you.

- In 2012 the world population reached 7 billion people.
- 255 births take place around the world every minute.
- That means that 4.3 births take place every second – can someone tell me how many new people have been added to our world during this assembly?
- The population is expected to reach 8 billion in 2025 – just 10 years away.
- In 1804 the world population reached 1 billion. It took 123 years to double to two billion. It took another 32 years to reach 3 billion. It took a further 15 years to reach 4 billion. It took 13 more years to reach 5 billion in 1987. What do you notice? What are the implications of this growth?

How will we feed, clothe, house, educate, care for all of those people? How will our earth cope? What measures will countries need to put in place to limit this growth and ensure that they can meet the needs of their people? Will there be even more conflicts as the competition for land and food and water becomes even greater? Certainly, it's something that we need to think about.

SUMMARY

We have learnt today about the changing map of our world. Sometimes the map of the world grows as we discover new places. Sometimes it shrinks as we discover new technologies and transport to take us to places more quickly. We have also reflected on globalisation and how this brings both problems and benefits to people and the environment. Looking to the future, we have considered how our world—especially the people part—is growing, and what consequences this may bring.

CLOSING:

Think about this quote from Jimmy Carter, former USA president:

> "Globalisation, as defined by rich people like us, is a very nice thing. You are talking about the Internet, you are

talking about cell phones, you are talking about computers. This doesn't affect two-thirds of the people of the world."

So, what does globalisation mean for them? We don't live in an equal world and the problems of population growth will only make it worse. One in two children live in poverty and one child dies every second around the world. So, whether our world is shrinking or growing, there could be problems ahead, and we must learn to steer our planet more safely and sustainably through these future times.

Thank you for listening.

* * *

AUTHORS

Mark Williams was Head of Science at St Georges School, Edgbaston, Birmingham. He now runs The Four Oaks Tutor Agency. He has written for Oxford University Press: The Association for Science Education, and was on the steering committee for the perspectives on Science AS level course. He has also written a lighthearted book about The Hash House Harriers.

Luke Citrine Williams has worked in both the British state sector and International sector as a teacher of Computer Science, Information Technology and General Science. In his previous roles he has also worked as a subject coordinator of Computer Science and Information Technology. He now lives in the United States where he teaches Middle School Science.

Donna-Lynn Shepherd is a psychology and education researcher at Coventry University, a school governor and a qualified secondary Maths teacher with special interests in Maths education, education in developing countries and conservation psychology. She has co-authored a number of articles for peer-reviewed academic journals as well as research reports for the Department for Education. Most of her research now takes place in rural Zambia in partnership with the African Lion and Environmental Research Trust (ALERT).

Juliet Stafford has considerable experience of working in secondary schools in the West Midlands over the past twenty five years. Her roles have included Head of Geography and Assistant Headteacher. She now works as an Education Consultant in local schools and has also been involved in examining, tutoring and writing educational materials.

Cover and illustrations by **Julie Hayes**.

Bibliography

1. Terry Fox assembly

   http://www.cancerresearchuk.org/home/

6. Making Trade Fair

   www.FairTrade.org

10. Successful People

    http://www.looktothestars.org/celebrity/richard-branson
    https://en.wikipedia.org/wiki/Anita_Roddick
    https://en.wikipedia.org/wiki/Alan_Sugar

14. Going for Gold

    www.brainyquote.com

17. The 3 Rs
    www.recyclezone.org.uk
    www.recylcing-guide.org.uk

Lightning Source UK Ltd.
Milton Keynes UK
UKOW04f0554030417
298192UK00011B/410/P